Documenting Racism

Documenting Racism

African Americans in US Department of Agriculture Documentaries, 1921–42

J. Emmett Winn

BLOOMSBURY
NEW YORK • LONDON • NEW DELHI • SYDNEY

Bloomsbury Academic
An imprint of Bloomsbury Publishing Inc

1385 Broadway
New York
NY 10018
USA

50 Bedford Square
London
WC1B 3DP
UK

www.bloomsbury.com

Bloomsbury is a registered trade mark of Bloomsbury Publishing Plc

First published 2012
Paperback edition first published 2013

Library of Congress Cataloging-in-Publication Data
Winn, J. Emmett (John Emmett), 1959–
Documenting racism: African Americans in US Department of Agriculture documentaries, 1921–1942/by J. Emmett Winn.
p. cm.
Includes bibliographical references and index.
ISBN-13: 978-0-8264-0555-5 (hardcover: alk. paper)
ISBN-10: 0-8264-0555-X (hardcover: alk. paper) 1. African Americans in motion pictures. 2. Racism in motion pictures. 3. Stereotypes (Social psychology) in motion pictures. 4. United States. Dept. of Agriculture. Motion Picture Service. 5. Documentary films—United States—History—20th century. I. Title.
PN1995.9.N4W56 2011
791.43'652996073–dc23
2011037680

ISBN: HB: 978-0-8264-0555-5
PB: 978-1-6235-6139-0

Typeset by Newgen Imaging Systems Pvt Ltd, Chennai, India

Contents

Preface

The maintenance and perpetuation of racist beliefs and values in the media ensures that the tenets of racist ideologies are broadly communicated and constantly reinforced. The communication of racist ideologies via the media is an important topic for cultural scholars because it is the site of hegemonic struggles over meaning. Cultural studies scholar Stuart Hall argues that "the media are not only a powerful source of ideas about race. They are also one place where these ideas are articulated, worked on, transformed, and elaborated."[1] Furthermore, media scholar Jane Rhodes has pointed out that a racist society needs the media to disseminate its principles.[2] Understanding how racism is encoded in media texts and communicated to audiences in both explicit and implicit ways helps scholars and media consumers to better understand the complex relationship between media and ideology.

Commercial filmmaking is an important site of study for anyone interested in the perpetuation of racist ideologies concerning African Americans, as negative black stereotypes have been utilized in filmmaking since its earliest years.[3] Racist representations of African Americans were appropriated from pre-cinematic culture and incorporated into commercial filmmaking from the beginning in films such as *Uncle Tom's Cabin* (1903), *The Confederate Spy* (1910), and *For Massa's Sake* (1911). The study of historic commercial films is central to understanding how racist ideologies were incorporated into the art and industry of filmmaking. For example, D. W. Griffith's 1915 film, *Birth of a Nation,* is often identified as both an important film for its impact on moviemaking and a bigoted film that perpetuates racist myths and stereotypes.[4] The use of negative stereotypes continued through the golden age of Hollywood in films such as *The Ghost Talks* (1929), *Stand Up and Cheer* (1934), and *Gone with the Wind* (1939) and has continued into the contemporary era in such movies as *The Color Purple* (1985) and *Mississippi Burning* (1988).[5] Indeed, Hall explains that versions of classic racist black stereotypes persist in the media today.[6] The continued interrogation of media texts both current and historic is important as scholars strive to understand and explain how racist beliefs and values are created and naturalized via communication in the media.[7]

However, in the early years of cinema not all motion picture producers were interested in the commercial aspects of film; many of them—for example, professional organizations, civic groups, local and state governing bodies, and the federal government—wanted to make movies for noncommercial purposes. These early noncommercial moviemakers saw the potential in film to communicate their messages to audiences who were eager to see this new source of entertainment and information. Furthermore, the power to move audiences emotionally through the filmic representation of events and stories appealed to noncommercial filmmakers who were intent on winning public support for their interests. For example, Stephen J. Ross describes how labor and trade unionists from 1907 to 1929 used motion pictures to "portray their cause visually."[8] Thus many groups with an interest in using film as a medium to encourage change and/or propaganda embraced filmmaking in its early years.

Eventually various political and socioeconomic factors collided to greatly curtail filmmaking by noncommercial filmmakers. However, several groups, including government agencies, continued to make movies for causes ranging from education to propaganda. The US Department of Agriculture (USDA) was one of the earliest agencies to produce films, and it did so for decades after the demise of many other noncommercial ventures.[9] Millions of people saw these free films in a wide variety of venues, including public schools, colleges and universities, civic meeting halls, libraries, church halls, and even open fields. Over the period of several decades, from the silent-movie era through the 1950s, the USDA was a significant government filmmaking organization, touting itself to be not only the first government agency to organize a filmmaking division but also the most prolific. The USDA motion picture branch was internationally respected, and the distribution of its films was high for US government institutions, particularly in the 1920s and 1930s. The USDA showed its films practically anywhere extension workers or others could gather a small group or a large crowd. In fact, for many Americans in the early 1900s, the USDA films were the first motion pictures they saw, and the experience of watching USDA-made films was shared by millions in the United States and around the world.[10] The role of the films varied, but the USDA saw all of its films as simultaneously entertaining audiences, in the broad sense of the term, while spreading the USDA's message and encouraging people to find out more about its work.[11]

Given the number of films that the USDA made, the size of its audiences, and the longevity of the USDA's Motion Picture Service, a surprising lack of scholarly attention has been devoted to the study of these films despite the fact that the department's "films on scientific and technical subjects, forestry and fire prevention, agricultural education and soil conservation have been a responsibility which has served the people well."[12] Even less research has been done on the representations of African Americans in Department of Agriculture films. For example, in his important book *The Negro in Films,* Peter Noble mentions only one of the USDA films devoted to African Americans, *Henry Browne, Farmer* (1942),[13] even though African Americans appeared in USDA films in the 1920s and 1930s as well. The USDA films made in the 1920s, 1930s, and 1940s that focus on blacks generally have escaped comprehensive analysis by scholars concerned with African American representations in film.[14]

Purpose and Scope

The current study addresses this situation by explicating the racist ideologies communicated through USDA motion pictures made from the 1920s through 1942. In this book I analyze the four major USDA motion pictures that focus on African Americans: *Helping Negroes to Become Better Farmers and Homemakers* (1921),[15] *The Negro Farmer* (1938), *Three Counties against Syphilis* (1938), and *Henry Browne, Farmer* (1942). My purpose is to explicate the representation of African Americans in these films within the racist ideologies of the period. Each analysis includes an in-depth investigation of the reality created by the filmmakers, the motion picture's use of stereotypes, the rhetorical message communicated, and a discussion of each film's relationship to broader racist ideologies. Finally, I build on the individual analyses to clarify how the motion pictures as a whole communicate the racist ideology that African Americans are inferior and a threat to white society.

It is important to note that this study is limited in its scope to the analysis of filmic texts and the supporting documentation available in the USDA archives and other sources. Thus, it is impossible for this scholarship to make broad general comments concerning the USDA or its programs, staff, or administrators beyond what the analysis supports. Moreover, the motion pictures often deal with specific institutions and historical figures, such as the Tuskegee Institute (now Tuskegee University) and a USDA extension agent named Thomas M. Campbell, because of their important work with rural African Americans. This study acknowledges the significance of such institutions and individuals and points out their accomplishments in relationship to the material presented in the motion pictures; however, it does not provide a comprehensive study or explanation of the important work that these institutions and individuals provided during the years these movies were made.[16]

USDA Motion Pictures, Documentary, and Reality

An important aspect of the analysis of these four films is an explanation of their constructed realities. In each of the films, the USDA uses conventions of documentary filmmaking to present a view of southern African Americans that it wants viewers to accept as accurate. These conventions encourage audiences to interpret the films as documenting reality. Some of the common conventions used are a narration that describes and comments on the action from an assumed position of authority; the implication that the scenes are of actual people and events (instead of carefully staged or scripted scenarios); and the use of actual locations, data, and statistics to convince viewers that the information is objectively presented by experts. Because they use these conventions, the USDA motion pictures are often referred to as documentaries.

Defining the USDA films as documentaries is problematic only if the reader equates *documentary* with the concept of an objective reality. The fact is that the USDA motion pictures analyzed in this book do not represent an objective reality of their times. To the contrary, they are carefully constructed to present a specific view of African Americans in the rural South that was approved by the USDA. That view, as I illuminate in my analyses, is informed and influenced by long-standing racist ideologies that defined southern blacks as inferior to whites and a threat to white society. The motion pictures do not overtly state this view of southern blacks but instead rhetorically argue such ideas through the use of consistent themes informed by racist ideologies. In other words, the USDA motion pictures are documentary films because they use documentary conventions, but they are not faithful presentations of an objective reality.

In essence, these motion pictures are constructed views of African Americans in the rural South during the decades from the early 1920s through the early 1940s. This concept of documentary film as a purposively constructed view of the world was not foreign to documentary filmmakers of the time. For example, Erik Barnouw, in his classic book *Documentary: A History of the Non-Fiction Film,* writes that movie producers understood that they could shape the viewer's understanding of a condition in the way they presented the situation in their films. In Barnouw's words, "The documentary film maker, dramatizing issues and their implications, could lead the citizen through the wilderness."[17] For example, Barnouw wrote that the famous documentary filmmaker John Grierson considered "cinema as a pulpit," to "not only explain" but also to inspire.[18] Likewise, in the first chapter of their 2009 book, *A New History of Documentary Film,* Jack C. Ellis and Betsy A. McLane investigate ways of understanding documentary and point out that "documentary is one of three basic creative modes in film, the other two being narrative fiction and experimental avant-garde."[19] From this position, they explain the common characteristics of documentaries, including subjects, purpose/viewpoint/approach, form, production method and technique, and audience response.[20] In discussing the second characteristic, Ellis and McLane explain that the purpose/viewpoint/approach "is what the filmmakers are trying to say about the subjects of their films. They record social and cultural phenomena they consider significant in order to inform us about those people, events, places, institutions, and problems."[21] Thus, documentary films should not be considered a filmic version of objective reality but a mode of filmmaking that allows the filmmaker's perspective to be expressed. The USDA filmmakers saw their role as not just educational but also propagandistic in the sense that they believed their films played an important role in helping the public to properly understand the USDA's programs and in eliciting support for the USDA's work.[22]

Finally, according to Bill Nichols, in his book *Representing Reality: Issues and Concepts in Documentary,* "Documentary represents the world of individual responsibility and social action, common sense and everyday reason. . . . This is indeed the world we see but it is also a world, or more exactly, a view of the world. It is not just any world but neither is it the only view possible of this one historical world."[23] Thus, it is important to remember that these USDA motion pictures present a view of the world that African Americans inhabited in the rural South during the 1920s, 1930s, and 1940s. However, that view should not be considered a presentation of the objective reality of that world, and in fact I argue that the view presented is biased by the racist ideologies of the times. There were then (and are now) other ways to view the world southern rural blacks inhabited in these decades that were not presented in these films. However, the USDA's view was privileged over other views because it was a powerful government department charged with an agricultural mission. This fact is also the reason why these films were widely distributed, trusted, and believed. Finally, the importance of the USDA's mission and its status as a government agency ensured that these films would be protected in the USDA archives for all time.

Analysis of the Films

Understanding that each of the USDA films creates a view of reality faced by rural African Americans in the early twentieth century is the foundation of the investigation into their ideological premises. In each motion picture, the USDA filmmakers carefully constructed a reality that was used as evidence of the government's work and the basis for the government's arguments. However, in working to craft a specific reality, the films drew upon the beliefs and values of the times. Scholars have demonstrated that ideological assumptions are ensconced in the media, myth, and culture.[24] Hence, the motion pictures perpetuate the ideological positions in which they are situated. Recognizing the important role that motion pictures play in the dissemination and maintenance of social beliefs and values, communication scholars have employed rhetorical strategies to interrogate filmic texts.[25] David Blakesley, in *The Terministic Screen: Rhetorical Perspectives on Film,* explains that this "approach views films as serving ideological functions in both its content and technical apparatus. The task of film criticism is to expose film's complicity with or deconstruction of dominant ideology."[26] Additionally, film scholar P. L. Simpson suggests that the way a film encodes ideological messages may "deflect attention from other possible" views of reality.[27] In order to understand films in their historical context, Thomas W. Benson suggests that a close textual reading of movies can shed light on how films combine form with public themes.[28] Moreover, Blakesley explains that "films reveal not only the predispositions of filmmakers, but they also serve ideological functions in the broader culture."[29] To this end, I analyzed these filmic texts in order to extricate the racist ideological suppositions ensconced in the texts and communicated to USDA audiences for decades.

My analysis of the films is supplemented by research from the USDA archives at the National Archives and Records Administration (NARA), in College Park, Maryland. The documents obtained for this research come primarily from the records of the office of the Secretary of

Agriculture, specifically "Records Relating to the History of the Motion Picture Service, 1908 to 1958"; "General Correspondence Relating to Negroes, 1909–1955"; and "General Correspondence of the Office of the Secretary, 1906–1970." These records provide a rich source of information about the USDA's motion picture efforts and valuable insights into its work with southern African Americans.

Chapter Outlines

Chapter 1 presents a brief history of the first four decades of the USDA Motion Picture Service. This introduction, along with this preface, orients the reader to the study and provides the background information required for the analyses of the motion pictures. Discussed in the following chapters are the four major productions concerning African Americans that the USDA distributed to both foreign and domestic audiences for several decades. The films communicated the USDA's view of African Americans living in the rural South.

In Chapter 2, I analyze *Helping Negroes to Become Better Farmers and Homemakers* and argue that the movie reproduced the racist separate-but-equal doctrine that ruled 1920s Alabama. The film denies African American farmers agency by relegating them to a position where they are unable to improve their farming conditions. Instead, the movie argues rhetorically that the power and ability to create positive change resides with the white USDA officials who direct the work of black extension agents. Thus, by interceding on behalf of black farmers, white authorities dramatically improved the lives of African American farmers and their families. Furthermore, Tuskegee Institute's pioneering work in black extension is acknowledged but downplayed, as credit for black agricultural outreach is transferred to the white authorities in the USDA. By cinematically presenting differences between blacks and whites, the film argues in favor of separate-but-equal laws and customs and communicates that blacks are inferior and require white help.[30]

In Chapter 3 *The Negro Farmer* is analyzed. *The Negro Farmer* presents several African American farm families who required assistance in most aspects of their farming and home lives, including nutrition, sanitation, gardening, livestock production, and crop improvement. The answer to all of these problems is presented in the context of the USDA's Live at Home program, which showed how, through extension demonstration work and 4-H club programs, these often dire situations could be improved. In doing so, the makers of the film chose to represent African Americans as inferior to whites and thus supported Jim Crow racism while opposing the migration of millions of African Americans from southern farms to the North and West. A key part of the film's opposition to the migration is its final segment, in which selected quotes from Booker T. Washington's 1899 essay "The Case of the Negro" are used to argue that African Americans belong on the southern farm, engaged in agricultural pursuits. By arguing that southern blacks should remain on the farm, the USDA's message also supported the racially oppressive system of segregation that forced southern African Americans into a state of economic peonage after slavery.[31]

Three Counties against Syphilis, the focus of Chapter 4, presents the story of the tri-county Public Health Service syphilis treatment and testing program for African Americans that utilized a mobile clinic to reach out to rural black communities in a sparsely populated area of costal Georgia. The significance of the film is heightened by the fact that the program and film were made during the early years of the infamous Tuskegee Syphilis Study conducted in Macon County, Alabama, in which hundreds of African American men were allowed to suffer the ravages of untreated syphilis. Unlike *Helping Negroes to Become Better Farmers and Homemakers* and *The Negro Farmer*, the title *Three Counties against Syphilis* does not overtly state in its title that its focus is on the African American community. However, after a brief period in the beginning of the film where it appears to be about a program that involves both whites and blacks, the motion picture forsakes its biracial approach to focus entirely on testing and treating African Americans. The film presents blacks as inferior to whites and argues that testing and treatment of blacks is essential to protect white society from black disease. The syphilis testing and treatment program was conceived and conducted within the premises of racial medicine, as was the Tuskegee Syphilis Study, a racist ideological concept that distinguished African Americans as a problem population, inferior to whites both physically and morally, and a threat to white society through social interaction. *Three Counties against Syphilis* communicated and perpetuated this racist ideology in the United States.

Chapter 5 is dedicated to my analysis of the 1942 World War II USDA motion picture *Henry Browne, Farmer*. This movie evaded representing blacks as dependent upon whites by locating this farm family in an all-black world where African Americans are completely segregated from whites. The movie tells the story of the Browne family, their farm in Macon County, Alabama, and the service of their oldest son in the all-black Tuskegee Airmen military squadron. The film presents the Browne family as successful and patriotic farmers who operate a model farm that employs the best agricultural practices. The motion picture demonstrates how the Brownes contribute positively to the US war effort by growing peanuts for oil, reusing materials, and using preventive maintenance to help prolong the use of farming equipment for the duration of the war. Henry Browne and his family are portrayed as patriotic American farmers who contribute positively. By segregating Henry Browne and his family and their accomplishments and service to the United States in an all-black world separated from white society, *Henry Browne, Farmer* rhetorically argues that Jim Crow, and its inherent racist ideology, was working positively for the United States and for African Americans. In other words, the film offers the impression that Henry Browne and other black farmers who were willing to remain in their place and be denied their fundamental civil rights could be successful and happy in the all-black world of Jim Crow segregation.

The final chapter, Chapter 6, uses the analyses of the four USDA films to demonstrate that the motion pictures communicated that African Americans were inferior to whites in all ways and threatened white superiority and white society, a fundamental belief of American racist ideology. The motion pictures accomplished this by communicating three broad themes (1) segregation is sacrosanct, (2) African Americans are farmers, and (3) blacks are a problem population. By consistently communicating the idea of black inferiority and the need for white supervision and control, the USDA films maintained the racist foundations of Jim Crow laws and customs, reinforced the doctrine of separate but equal that denied African Americans their civil rights, and upheld segregation not only as the natural condition for blacks in the United States but also as the best situation for blacks to succeed in America.

Acknowledgments

In researching and writing this book, I benefited from the help, encouragement, and support of many wonderful people who, mostly because of their goodwill, unselfishly aided me in a variety of ways. To each of them, I owe a public thanks and am happy to attempt to pay that debt with these lines.

I thank my parents, Albert Taylor and Roger Lee (Gandy) Winn, for all of their support and encouragement through the process of producing this book and for all the help they have given me for the last half century. Roger Lee and Albert have always worked hard and taken pride in their work, regardless of the task at hand, and thus were terrific role models. I am grateful for their love, lessons, and support.

I am very fortunate to have the support of a wonderful group of friends and colleagues who offered encouragement and assistance in a variety of forms, including good company, great food, and even better advice. These fine folks include John and Kathy Tamblyn, Robert and Susie Weigel, Rob Agne, Tim White, David Payne, Danni Williams, Julie Huff, and Kim Golombisky. I'm thankful that they are always there, even when I am not the best of company.

The nature of this book required a good deal of work in archives and libraries, and I am indebted to the many archivists and librarians who helped me. I am particularly grateful to archivists at the National Archives and Records Administration II, in College Park, Maryland, especially Vernon Early, an archive specialist who went well above and beyond, on short notice, to produce a synchronized sound-track copy of one of the films for my study. Additionally, I thank Dean Bonnie MacEwan and all of the wonderful staff and faculty of Auburn University's Ralph Brown Draughon Library for their assistance.

I have benefited greatly from the help of three professional editors. My sincere thanks go to Jill R. Hughes of the Editor's Mark, who has provided encouragement over the last year as well as excellent copyediting and proofreading. I owe a great debt to David Barker, editorial director at Continuum, who was instrumental in aiding me with the proposal and subsequent planning of this book. Finally, it has been a pleasure working with Katie Gallof, Continuum's Film and Media Studies acquisitions editor. I cannot thank her enough for her encouragement, help, and humor as we thought through ways to make the book better during the writing process.

Finally, and most importantly, I thank Susan Brinson, my spouse, friend, and colleague. Susan provided invaluable intellectual insight and emotional support through the entire process. I am very grateful to share my life and my work with Susan.

Chapter 1

A Brief History of the USDA Motion Picture Service to 1943

An understanding of the history of the USDA's Motion Picture Service (MPS) is helpful in placing the USDA films that focused on African Americans in the broader context of the USDA's motion picture program.[1] The history of the MPS demonstrates that it took its work very seriously and strove to produce quality films within its limited budget and resources. The MPS grew from a small group of enthusiasts within the USDA to a major division that produced hundreds of films over a period of several decades. In its earliest years the MPS operated on a level that was comparable to most film production, even in the commercial realm, but soon settled into a respected position in the governmental film arena.

There is no doubt that the USDA began producing films very early relative to other government agencies. In fact, the early MPS pioneers claimed that the USDA was the first government agency to produce a motion picture. For example, an MPS unit chief reported that in 1908, "when the Wright brothers came to Fort Myer [Fort Myers, Florida] to demonstrate their pioneer biplane for the Signal Corps, the Department of Agriculture made a film record of the demonstration—the first government motion picture."[2] Others have questioned this assertion; for instance, in their 1923 book, *Motion Pictures in Education,* Don Carlos Ellis and Laura Thornborough claim that the US Reclamation Service was the first "among the bureaus in Washington to take up this work, and . . . exhibited at the Jamestown Exposition 1907 films showing the work of the Government in reclaiming arid lands."[3] However, the right to the claim of being the first government bureau to make a motion picture is of less importance than the fact that the USDA very quickly became a leader in government-sponsored motion pictures and was considered a model of success in the endeavor.[4]

The earliest motion picture work of the USDA was not part of its official duties and was performed without official departmental sanction. Ellis and Thornborough contend that the first USDA staff member to make a government film was Joseph Abel, a Bureau of Animal Industry photographer, who in 1909 "made a film of the live-stock show at the Alaska-Yukon Pacific Exposition and, later, films on the Texas Fever Tick and Its Eradication."[5] These early efforts were concealed from USDA officials. According to the MPS staff, a film laboratory was established without the knowledge of the secretary of agriculture. The reason cited was that the USDA secretary at the time, "Tama Jim" Wilson, thought the "motion picture was the work of the devil."[6] This situation allegedly changed when O. H. Benson, a pioneer in the 4-H Club movement, surreptitiously set up a motion picture camera and filmed Secretary Wilson speaking to a group of "corn club boys." When the secretary saw the footage, "he was amazed and delighted—and from that moment the motion picture program of the Department ceased to be a bootleg enterprise."[7] Some reports cite C. Francis Jenkins as the camera person for this footage.[8] Jenkins was a government clerk who later became an important film pioneer.[9] Regardless of the exact reason for the legitimizing of filmmaking, the USDA film work continued in an ad hoc manner for several more years.

A modest film laboratory was established in 1912 in the USDA Section of Illustrations, where staff employees W. S. Clime and G. R. Goergens "were regularly assigned motion picture work."[10] The USDA and others have identified this as "the first Government motion picture laboratory to be established in the world."[11] Although the USDA staff members took such bragging rights very seriously, there is little historical evidence to confirm or deny these claims beyond the statements of the staff members in their written reports. However, the fact that the early USDA motion picture production was impressive and well respected in the educational film arena is not disputed. For example, Ellis and Thornborough argue that the USDA films had "considerable educational value" and "constituted the largest single collection of educational films on a related group of subjects."[12] This success in the early years, spanning from about 1908 to 1912, brought the USDA positive attention in the burgeoning film industry, both commercial and noncommercial.

The USDA's efforts were so successful that in September 1913 the agency formed "a temporary committee to determine whether the Department should enter the educational motion picture field in earnest."[13] Later that year the USDA established a permanent committee and authorized the Division of Publications (Section of Illustrations) to "purchase and install a complete motion picture photographic apparatus."[14] It was charged "to install a Motion Picture Laboratory, to purchase the necessary apparatus, and to produce motion pictures for visual instruction in agriculture under the direction of the motion picture committee."[15] Thus, in late 1913, the earliest official USDA motion picture unit was established. By most measures, the USDA's task was daunting, given that the USDA at that time consisted of seventeen different bureaus with wide-ranging responsibilities. Ellis and Thornborough cite this diversity as a main reason for the USDA's success in the government film arena, because all of its motion picture work was coordinated by a single film unit, thus athe MPS was able to produce movies on a wide variety of topics.[16] This variety tested the USDA filmmakers while also providing them with ample opportunities to experiment with new techniques and ideas.

Establishing what amounted to a small but fully functioning movie production studio within the federal government in early 1914 proved to be a challenge for the USDA staff. Early reports to the USDA secretary indicate that the MPS was experiencing problems and asking for patience. Since the MPS had a very small budget, the staff had to carefully research motion picture equipment before deciding what to purchase. For example, they performed a fairly extensive study of lighting to decide what would work best for their films.[17] Moreover, faced with the problem of exhibiting their films in rural areas, the staff needed to understand the challenges of exhibition in a noncommercial environment.[18]

Eventually the MPS settled into a production strategy that began with the creation and approval of a basic scenario that mapped the subject matter and shots for the motion picture.[19] Recording the images was simple at first; static camera setups filmed short vignettes, which were explained or augmented by the use of intertitles. The editing was simple and the films were generally short (usually one to four reels of film). As the MPS library of films grew over the years, it was necessary to revise some motion pictures "due to changes in agricultural methods and practices," and often films would need to be "withdrawn from circulation as they become obsolete."[20]

Distribution of the USDA movies was accomplished through two main methods. The primary distribution mode was lending the films, without charge beyond shipping, to the state USDA offices (very often to state extension agents) and other governmental, educational, and civic groups. The distribution was originally, and for many years, handled from the Washington, DC, office, which required that statistics on the audiences be returned with the motion pictures. The popularity of the movies soon surpassed the small number of film prints that the MPS budget allowed, and it became normal practice for the USDA to refuse hundreds of requests per year. The second method of distribution was direct purchase of film prints. For many years the USDA provided various ways for outside groups and foreign governments to purchase copies of their films at a nominal price relative to the cost of producing the prints. Sometimes the MPS made these copies in-house, but the most common method was to contract with a commercial firm to make prints directly for the buyers.

Film exhibition was accomplished in both commercial and noncommercial venues. Often the recipients of the movies, such as the state USDA agents, showed them in the field as part of a larger demonstration event.[21] Several state USDA offices employed trucks outfitted with generators and exhibition equipment so that presentations could be made in any area, no matter how remote.[22] Common public locations such as schools, libraries, churches, and social halls were also used to exhibit the films. In addition, USDA films often enjoyed commercial exhibition in movie theaters, where they were presented with "news weeklies" and other short, magazine-type nonfeature programming that was common at the time.[23] USDA films were also often shown as part of the war effort during World War I and World War II; for example, films promoting the backyard poultry movement were shown during World War I.[24]

Early on, the USDA learned that showing its motion pictures at USDA extension personnel presentations in rural communities drew larger audiences than talks without motion pictures. A November 1914 report explained that "the announcement that the motion picture exhibit was to be a feature of the lecture, attracted large audiences, even in bad weather and under difficult road conditions. . . . In every case, the motion picture exhibit brought out 75 to 100 people, where stereopticon talks attracted 10 to 20."[25] This result was consistently repeated, and it soon became obvious to the USDA that a major benefit of the films was in drawing larger crowds to its local meetings. The USDA saw this growing demand for its motion pictures at the same time that it was struggling to produce new films.

This led the USDA to consider allowing commercial film production companies to produce its motion pictures. Several film studios showed interest, including the Historical Film Company of New York and London, which offered to make films "of agricultural methods and all other ramifications to be used not only as records in the department in Washington, but for [the] department's missionary work throughout the country."[26] Although the USDA did not contract with the Historical Film Company, the department maintained a keen interest in commercial collaborations.

One successful USDA collaboration was with New York's Universal Film Manufacturing Company. After a USDA staff member resigned from the MPS to accept work with Universal, a USDA memorandum was issued that urged cooperation with the company, as it planned to make "sets of educational films illustrating the work of the Department." The memo indicated that this was "an opportunity for legitimate publicity" for the USDA.[27] The USDA was interested in partnering with commercial film companies as a way to help stretch its small motion picture budget. Hence, USDA staff visited commercial film companies in New York in order to "determine whether we cannot enter into some valuable cooperative arrangement with the commercial houses."[28] The USDA ultimately decided that the only way it could control the content of its motion pictures was to produce them in-house, as "it soon became apparent that the Department could not . . . establish adequate censorship over its own subjects or make satisfactory financial arrangements covering the purchase of films" with the commercial houses; so the USDA abandoned the idea of allowing commercial interests to make all of the department's motion pictures.[29]

By 1917 the MPS staff reported that motion picture production was a staple of the USDA's work.[30] Following US involvement in World War I, the MPS made films for the war effort "under the direction of the Committee on Public Information."[31] Many of the films addressed issues related to raising backyard poultry and home canning and preserving of foods.[32] However, the largest USDA motion picture campaign in support of the war "dealt with farm labor, making appeals for the enlisting of city people in farm work."[33] The USDA believed this promotion was very successful and reported that "a direct result" of the films was that many farm laborers were recruited.[34]

Even though the USDA had decided to produce the bulk of its films in order to maintain strict control over content, the department did not avoid all commercial collaborations. Interested in having its films distributed widely, the USDA approached several companies, including "Artcraft, Equitable, Metro, World, Brady, Fox, Triangle, Universal, Vitagraph, Edison, V.L.S.E., Ultra, Chester, Prizma, General, Paramount, Mutual, Pathe, International, Hoffman, Selznick, Educational, Argus, Gaumont, and Bluebird,"[35] about distributing USDA films in commercial theaters. In 1918 the USDA entered into an agreement with the Universal Film Company.[36] Universal screened several USDA films in commercial theaters, often as part of Universal's Screen Magazine series, with the first USDA film appearing in *Universal Screen Magazine No. 47* with "subsequent releases . . . at least bi-monthly thereafter."[37] This agreement proved very successful; a 1918 USDA report estimates that four million people saw the USDA films exhibited in commercial theaters.[38] The USDA was so pleased with these results that it had a filmic version of the USDA seal created for use in motion picture titles, thus improving their promotional value to the department.[39] Despite its success, the USDA did not continue to contract exclusively with Universal for long but looked for ways to collaborate with other commercial entities. Consequently, its films continued to be shown in commercial theaters, in one manner or another, for several decades.

One particularly successful collaboration with a commercial film studio occurred in 1919, when the USDA purchased an animated film from Bray Studios titled *The Charge of the Tick Brigade.*[40] The USDA avoided its concern about content control by approving the motion picture's scenario before allowing Bray Studios to produce the movie. The result was the USDA's first animated film, which proved to be a huge success with audiences. Years later, Raymond Evans, chief of the MPS, wrote that this "one-reel cartoon . . . was the spear-head of a long and arduous campaign that finally resulted in the eradication of the cattle fever tick from practically the whole continental United States."[41] *The Charge of the Tick Brigade* was an excellent example of a successful collaboration between the USDA and a commercial studio.

Interestingly, *The Charge of the Tick Brigade* and a related live-action film titled *Mollie of Pine Grove Vat,* also about tick eradication, may have been among the first motion pictures seen by some rural southern audiences. Years later a report from an MPS staff member claimed that "thousands of persons . . . in the hinterland districts of the South will remember this [*Mollie of Pine Grove Vat*] as the first motion picture they ever saw, as the . . . motion picture truck carried [it] to many remote communities where commercial films had never been seen."[42] The USDA considered these two motion pictures crucial to the success of the tick eradication effort because it believed that the films had been able to win over opponents of the USDA's campaign.[43] For instance, one report described how a USDA movie truck driver and tick eradication agent, Ed F. Pickering, had shown *Mollie of Pine Grove Vat* to "an audience that included a crowd of bullies and the local 'bad man'" in a community hostile to tick eradication efforts, where "threats had been made to blow up the 'damn tick wagon.'" The film was so successful in promoting the USDA's work that after the movie, the "leader . . . shook hands with him [Pickering] before he left."

The MPS had achieved a good level of production for its resources by 1920. The previous year, they had completed twenty-six reels of film on eighteen different subjects and considered its production "greatly increased" over the earlier years' efforts. Moreover, the MPS reported that "the quality of films has been noticeably improved,"[44] a particularly noteworthy comment since the MPS staff was sometimes criticized by others in the department for inconsistent quality in their films.[45] Distribution concerns also were a problem for the MPS, as their budget did not allow them to make many prints of their motion pictures. However, 1920 also saw an improvement in this area when the department arranged for outside institutions and individuals to purchase copies of the motion pictures, thus relieving some of the budget concerns and resulting in what the MPS considered a "substantial increase in distribution of those films."[46] By this time the MPS was fairly well established in the production and distribution of its films, and the USDA saw value in this work.

As a result of its success, in 1922 the USDA constructed a new film lab for the MPS with better equipment; these improvements raised the USDA's facilities for filmmaking to a level "equal to those of any other producer of educational films."[47] The MPS believed that these improvements, along with its formal recognition as a distinct and separate unit reporting to the director of extension, demonstrated that motion pictures had become established as important to the USDA's "extension, educational, and publicity work."[48] As a result, the MPS completed twenty-eight new movies and revised forty-five older films in its library that year.[49] This production brought the total MPS film collection to 169 subjects; 1,047 prints; and 1,237 reels of film.[50] Moreover, between 1922 and 1923 the MPS reported an increase in audience attendance of approximately 2 million viewers, for a total of 4.46 million; it suggested, however that "allowance should be made for possible exaggeration, but this consideration is balanced by the fact that many users failed to report their showings. In addition, there are no figures available in regard to the exact size of the audiences reached by the department films that have been bought by cooperating or outside institutions."[51] Regardless of whether 4.46 million viewers is an accurate estimate, it is clear that by 1923 the USDA's motion pictures were reaching millions of viewers per year. It should be noted that audience members were not just noncommercial US audiences; the MPS 1923 annual report explains that portions of their films continued to be included on occasion in the commercial "news weeklies" distributed to commercial audiences and that some of their films were exhibited in Europe and at the Brazilian exposition in Rio de Janeiro.[52]

Film work at the MPS remained steady through the end of the 1920s. For example, the 1927 annual report explains that the MPS completed 18 new motion pictures, bringing the total of MPS films to 222 subjects in 1,792 positive copies for loan to a reported attendance of 4,300,864. USDA movie sales also continued, with 200 prints sold in 1927, including 80 purchased by ten foreign countries.[53] The 1928 annual report gives a different indication of the distribution success of the USDA films, stating that the USDA made approximately 8,000 shipments of its motion pictures that year "for a total loan period of nearly 100,000 film days." According to the report, the USDA had to turn down 600 other requests "because of lack of copies to meet the demand."[54]

Prior to the end of the 1920s, the USDA films were available only in 35-millimeter format; however, by 1929, individuals and institutions could purchase USDA films in 16-millimeter,[55] and within a year the USDA began limited circulation in 16 mm.[56] Sales of USDA films (in both 16 and 35 mm) continued at a consistent rate, with domestic sales being led by agricultural colleges, first and foremost, and by other higher education institutions.[57] Foreign countries also continued to be very interested in purchasing USDA films; 1929's international buyers included Mexico, Australia, Japan, India, Turkey, New South Wales, Russia, and South Africa, among others.[58] Additionally, the MPS produced a film titled *Naturalized Plant Immigrants* for the international exposition held in Seville, Spain, and provided copies of twenty-five of their existing films to also be shown.[59] Hence, by 1929, thanks to this success, the demand for films far exceeded the MPS's ability to provide prints to its audiences by its traditional loan method. In addition, a technological revolution was occurring in the film industry, and so in 1929 the MPS began experimenting with sound pictures and soon realized that it would have to retool its production.[60]

An MPS staffer proclaimed the silent film "dead" in 1930.[61] USDA filmmakers realized that "commercial developments made it possible to place sound projectors on the market at a price within the reach of users of Department films and there was a sudden demand for the 'talking' picture."[62] MPS filmmakers were concerned because they realized that the costs of sound production were well beyond their resources and facilities; however, they understood that several USDA units were very interested in the educational possibilities sound films might offer, particularly in the idea of a "canned lecture" that could be "illustrated with motion pictures . . . [as a] method of teaching."[63] Therefore the MPS began working toward the goal of producing sound films.

The issue of exhibiting sound pictures in the field was a concern for the USDA, and early on, the MPS suggested that state extension officers "provide projection trucks equipped for putting on good talking picture shows, and route them about the states in accordance with local demands."[64] Moreover, in a further move away from 35-mm as the standard for USDA film exhibition, the MPS suggested the narrower width, because the 16 mm sound-on-disc equipment was easier to handle, and argued that "the same discs used with the 35 mm. prints can be used with the 16 mm. prints."[65] Overall, the MPS staff believed that local extension agents could better handle the 16 mm sound format in the field and preferred the sound-on-disc technologies to the sound-on-film technologies available at that time.[66]

In terms of production, the MPS realized that they had to immediately reimagine their filming practices, with one staffer saying, "The sudden advent of the 'talking' motion picture has necessitated a complete revision of the type of picture produced. No longer is it practicable to photograph persons in the act of talking so that lip movement is visible. All pictures which suggest an accompaniment of sounds are deleted, or so assembled that an accompaniment of sound may be furnished at a later date."[67] After reviewing both sound-on-disc and sound-on-film technologies, the MPS purchased disc sound projection equipment and began working on synchronized sound productions.[68]

By 1931 the MPS staff had established a mature film production studio and had transcended the act of simply manufacturing motion pictures; instead they were now dedicated to "develop[ing] a [film] laboratory along experimental lines, so that it might make a contribution to the motion picture art, especially as applied to educational work."[69] To that end, an MPS film engineer developed a device to allow for time-lapse photography, which would be useful in filming such subjects as seed germination.[70] The MPS also progressed well in its experimentations with sound projection and recording equipment and made some investigations into color film production.[71] Despite these advancements, the USDA saw a decrease in demand for its movies of approximately 15 percent due "largely to the fact that the demand for silent pictures [had] tended to abate since the advent of talking pictures."[72]

Nevertheless, the MPS continued the retooling to sound production and struggled to begin producing more sound pictures to meet the demand for "talking" movies.

The early 1930s saw the MPS grow to a staff of twenty-five permanent employees.[73] The 16-mm format continued to grow in popularity because of its "ease of handling."[74] Complaints from borrowers about the condition of some of the movie prints prompted the MPS to institute a more rigorous system for inspecting the prints before they were shipped in order to remove from circulation prints that were badly marred. This system resulted in the USDA having fewer bookings but increased the satisfaction of borrowers.[75] Sales of prints remained strong, with a total of 244 films sold in 1932 (45 of these purchased by foreign governments).[76]

The year 1932 saw the release of the first sound motion picture produced by the department, *An International Study of American Roads.* Although the sound is mostly a postproduction musical score played by the US Marine Band, the film included "an introductory speech . . . [by] Thomas H. MacDonald, Chief of Bureau of Public Roads of the Department."[77] The bulk of the motion picture was shot silently; however, the synchronized sound used on the speech was an important advance for the MPS in realizing their ambitions to produce "talking movies." Thus, by the end of 1932, the MPS could simultaneously record both images and sound in the studio and project sound movies in their facility using portable sound equipment.[78] By 1934 the MPS had a total of 4,270 reels of film in its offerings, with 269 of them containing synchronized sound.[79]

The MPS staff's efforts were rewarded in 1935, when they were relocated into a new film production lab in the newly constructed Department of Agriculture South Building, which touted what one MPS staffer described as "the largest and best equipped [film lab] in the government service, including a commodious and well-appointed sound stage."[80] The MPS settled into their new facilities and worked to regain the losses they had experienced with their delayed change to sound films. By 1938 the MPS was once again reporting annual audiences in excess of 5 million and routinely being forced to deny hundreds of requests per year due to a lack of prints.[81] To help with this situation, the USDA established cooperative agreements with institutions, often state universities, to aid in handling the distribution of films to their users.[82] In addition, the MPS, cooperating with the Civilian Conservation Corps (CCC), provided the organization with fifteen hundred prints of USDA motion pictures to be used for educational activities. Through this agreement the MPS reported that more than ninety-nine thousand CCC "enrollees per month were given agricultural film study periods."[83] Thus, the USDA's film audiences had rebounded and grown with the addition of sound motion pictures to their offerings.

As the 1930s ended, the MPS reported that it had reached its limit with three thousand movie prints on approximately two hundred different subjects. In a statement the MPS staff explained that "this is about the maximum number of subjects that can be maintained, the annual discards now offsetting the annual accretions."[84] Additionally, the MPS was unable to produce more than fifteen to twenty-five new films per year because of their limited resources.[85]

The early 1940s held many changes for the MPS. The sale of copies of their films was handled via annual contracts with commercial laboratories, and several hundred were ordered each year.[86] USDA films continued to be screened to millions of viewers, and the MPS began experimenting with Spanish-language versions of some of their films. For example, in 1942 the MPS began preparing a Spanish translation of *The Battle Is in Our Hands,* with Andrew Cordova as the Spanish narrator. Additionally that year, a Spanish-language version of *Democracy at Work in Rural Puerto Rico* was in production.[87]

However, the largest change occurred in 1943, when the MPS studio, the film lab, and fourteen staff members were transferred to the Office of the Coordinator of Information (later the Office of Strategic Services, OSS) to make films related to the war effort. The MPS reported that they were left "with a skeleton staff, inadequate production equipment and no suitable working space. However, we did have some cameras, editing equipment and trucks and with this as a nucleus we rented a building vacated by the Paramount Exchange in Washington and [continued] with our production."[88] Despite their reduced resources, the MPS continued to produce several films per year by contracting out sound, animation, and photographic lab work.[89] In December 1945 the borrowed staff and facilities were returned from the OSS to the USDA. The facilities included the "motion-picture laboratory, cutting rooms, storage vaults, projection rooms, sound stage and studio, animation studios, and other facilities."[90] The MPS reported that their unit was "the only civilian Government agency in Washington with complete facilities for motion picture production."[91] The MPS continued to make motion pictures for the USDA and other government agencies for several years.

As radio and television played a more central role in the media, the MPS's role diminished until an in-house USDA film production unit was no longer viable. The NARA archives devoted to the records of the MPS end in 1958; it appears that the MPS was functionally inactive by that time. However, the history of the MPS is important to everyone interested in governmental and noncommercial film production. The USDA's efforts began during the early years of cinema and continued well into the era of television supremacy in the US media.

Chapter 2

Helping Negroes to Become Better Farmers and Homemakers

The US Department of Agriculture produced its first movie focusing on African Americans in 1921. *Helping Negroes to Become Better Farmers and Homemakers* relates the efforts of county extension agents in eastern Alabama to teach African Americans how to build better farms and make better homes. The film was part of a larger USDA effort in extension demonstration work that focused on helping rural Alabama farmers improve their living conditions and their crop yields through advances in scientific farming and tested business models. The project also included photography, the production of written materials, and (several years later) the addition of radio broadcasts.

The film's events center on the members of an African American farm family named Collins as they are being taught to become better farmers and homemakers by their extension agents. Thus, the motion picture is a dramatization of the types of outreach work that the agents attempted in their communities. Although the events are dramatized and staged for the movie, it is reasonable to assume that the people represented in the film are actual farmers, county agents, and administrators. Many of the film's scenes show African Americans living on the farm and their participation in the extension service's demonstration work in their community. There are scenes of cotton farming and home improvement activities that have historic appeal for the twenty-first-century viewer. Similarly, the motion picture includes scenes of the Tuskegee Institute campus from the early 1920s and scenes of the mobile school that Tuskegee maintained as an important form of outreach to African Americans farmers in Alabama.

The film's stated objective is to show the audience how the USDA's extension work helped rural southern African American families improve their farming methods and create better living conditions through extension service demonstration work. The extension service is a national network of county extension agents created by the Smith-Lever Act of 1914 and headquartered at land-grant universities. In Alabama the white land-grant university was Alabama Polytechnic Institute (API), located in the city of Auburn in Lee County (API changed its name to Auburn University in 1960). Tuskegee Institute, in neighboring Macon County, was a center of black extension work. Tuskegee was founded by the Alabama state legislature in 1881, but by 1883 it was a private institution. The choice to use Tuskegee as a black extension center was made because of its long commitment to agricultural outreach and instruction. Its founder, Booker T. Washington, was devoted to reaching out to black farmers, and he both initiated and developed the basis of black extension work.[1]

Given the use of racist stereotypes in motion pictures in the 1920s, it is not surprising that this film perpetuates these stereotypes in support of the prevailing cultural and political practice of segregation and belief in black inferiority. In so doing, *Helping Negroes*, as a cultural artifact documenting institutionalized racism, exposes the government's rhetorical support for the doctrine of racial separation. But the film goes further by systematically locating the agent of change to the dominant white population while simultaneously supporting the view that blacks were inferior.

Reasons for Making *Helping Negroes to Become Better Farmers and Homemakers*

The USDA archives do not contain documents that give the specific reasons for this USDA undertaking in the early 1920s. However, the archival material suggests at least three reasons. First, the USDA MPS received many requests from various divisions within the USDA and from other governmental agencies to make films highlighting their programs and work because the films received positive attention and were successful at drawing crowds to programs across the country. The extension agents were enthusiastic users of the USDA's films and informed the department that they experienced much better turnouts at their meetings when motion pictures were part of the programs.[2] Given that *Helping Negroes* spotlights extension work, the extension service would have been keen to have this film made.

Secondly, for years prior to 1921 the USDA received many questions about its efforts to help African Americans, requests to offer more help, and sometimes, admonishments for not doing more to aid black Americans, especially in the South. It was obvious to many that southern blacks needed help. Historian Jack Temple Kirby reminds us that "the typical black farmer . . . especially in the lower South plantation states . . . was a sharecropper."[3] Sharecropping was a form of tenant farming where the tenant farmer lived on the landlord's property and farmed part of the land for a share in the crop's revenue; "essentially it was a form of debt peonage."[4] According to historian David Eugene Conrad, "Southern tenancy was a vicious self-perpetuating system. . . . the workers were little more than serfs, held to the land by debt, ignorance, poverty, and dependence on the landlord."[5] The USDA's response to these requests and concerns was to point to its use of black extension agents in the South to show that it was both concerned about black Americans and doing all that it could to help the black community. From the USDA's perspective, the black extension program demonstrated its commitment to African Americans because it not only employed blacks as extension agents but also helped blacks better their lives through improved farming techniques. The USDA often cited the number of black extension agents it employed as proof of its work.[6]

Third, *Helping Negroes* appeared at the same time that a 1920 USDA circular titled "Extension Work among Negroes, 1920" was issued.[7] One section of the circular refers to the purpose of *Helping Negroes*: "to show step by step the development and progress of a typical Negro community under the influence of the demonstration work. The scene is staged in Alabama, where Negro extension work was first started. This film gives a striking insight into real conditions and how they are being improved in many sections of the South, and should be of great educational value."[8] The circular explains that "this circular shows how this work for Negroes, conducted by agents of their own race under the supervision of the agricultural colleges and the US Department of Agriculture, is helping to make the farm life of this great body of citizens more profitable and attractive."[9] The idea of making farm life more attractive to black Americans is better understood in terms of the migration of blacks from the rural South to the industrialized North, also referenced in the circular as "a very serious problem."[10] Kirby explains that in the 1920s and 1930s tenant landlords became frustrated as African Americans migrated away from southern farms because the landlords needed their cheap labor.[11] At least part of the point of the extension work and the circular was to demonstrate how the USDA was helping to address the problem of blacks leaving the farm for the city. The USDA appears to have decided to produce Circular 190 and the motion picture *Helping Negroes* to publically tout its efforts to help black Americans to remain in the South as tenant farmers.

Cinematic Representation versus Historical Evidence in *Helping Negroes to Become Better Farmers and Homemakers*

Helping Negroes is a thirty-two minute silent film consisting of a series of approximately seventy-three intertitles that introduce or explicate an equal number of vignettes; the very short intertitle/scene combinations average about forty-three seconds each. Most of the shots are composed of single camera setups; only a few of the scenes are constructed by a group of shots edited together or utilizing mobile camera techniques such as panning or tilting. The mixture of documentary-style scenes and dramatized events, explained or interpreted by the corresponding intertitles, tell the main story of the film: how a black farm family's life is transformed through the help of extension work.

The first intertitle explains that the film was produced by the US Department of Agriculture Educational Film Service. The term *educational*, as it is used here, does not mean instructional in the sense of teaching farmers particular skills or practices. The Department of Agriculture realized very early on that the primary benefits of their films were "attracting larger audiences, the pleasure the people derived from seeing the pictures, and the aid to complete understanding of the lecture or subject that the picture afforded. The pictures enabled those who saw them to visualize into concrete action the otherwise abstract points of the propaganda."[12] In fact, the film offers little in terms of actual instruction.

Reconstructing Reality through Content and Form

The bulk of the film revolves around a family of black tenant farmers, Rube and Hannah Collins and their children. The audience is introduced to the family first by an intertitle exposition, followed by a scene depicting that description. For example, the film's third intertitle reads, "Rube Collins, typical of many Negro tenant farmers in the South." A scene of an older black man hammering out a metal farm implement on an anvil follows. The audience is introduced in the same way to Rube's wife, Hannah; one of their sons, Obe; and one of their daughters, Ca'line. Likewise, the Collins home is presented. It consists of two unpainted buildings with no windows. One is a single-room cabin with a chimney, and the other is a poorly constructed rough-hewn one-room shack. (The representation of these types of shacks and cabins as African American farm homes continues throughout three of the films analyzed in this book.) These two small structures provide a visual cue to the audience that the family is poor and needs assistance from the USDA. In describing the deplorable conditions of a typical sharecropper's home, Conrad writes that "wind and rain came through holes in the roof and walls. Plumbing was unheard of and outhouses were considered a luxury."[13] According to the movie, the Collins family produces cotton, raises chickens, and owns a mule. These scenes are intended to demonstrate the poverty of a typical black farm family in Alabama at the time. This choice on the part of the filmmakers to show the disrepair of the family's meager material possessions sets up the visually striking change that will occur on the Collins farm after the intervention of the USDA agents. That conversion becomes rhetorical proof of the power of the white agents to enact change.

The story of the USDA's assistance begins when a neighbor tells Rube about a pest he has found in his cotton crop. Collins checks his field and also finds the pest, and the two men ride their mules to town. The intertitle explains that they are taking the news of the pest to their landlord. When they arrive in town, their white landlord summons the white county agent. The agent tells the two farmers that they have boll weevils and explains a way to attempt to control the pest. The boll weevil is a beetle that migrated to the United States in the late nineteenth century and threatened the southern US cotton crop by eating the cotton. In the early 1920s, when this film was made, the pest was destroying the lives of southern cotton farmers by ruining their cash crop.

The three scenes that involve attacking the boll weevil infestation show the farmers executing the county agent's instructions. These are the only instructional vignettes in the movie, but they do not offer any specific instruction to the viewer. This is a common problem in USDA films, and as extension expert A. I. Tannous says, much of the "instructional value is lost because they [the moving pictures] move too fast."[14] However, this sequence served a rhetorical function by showing how the county agent improves the black farmers' lives. The film states that "following the agent's instructions and with their landlord's approval Collins and his fellow tenant farmers begin to diversify their farming and to grow food and feed crops." This is a significant shift, because it ignores Rube Collins and renders him passive to the white landlord and the white extension agent. However, it is misleading for the filmmakers to frame this improvement in black farming methods as attributable only to white agents. Historically, it is accurate to assume that white extension agents helped black tenant farmers in the South, but farm outreach to black farmers in Alabama was carried out by black instructors from Tuskegee before the extension service was involved and later by black extension agents.[15] The filmmakers' choice to ignore Tuskegee's work prior to the USDA's formation of the black extension program may have resulted from the fact that the agency was making the film to highlight its programs. The choice also establishes early on that the source of power and knowledge for improving the lives of the African American farmers lies in white authority figures, not in the black community.

The next intertitle states another misleading point: "Two years later, better farming has given the Collins family a better home, and a local Negro agent *to assist the white agent* has become *necessary*" (my emphasis). The scene shows the Collins family working outside their much larger farmhouse, which contains several rooms and many windows. The Collins family's fortunes have improved, and the film attributes these achievements to assistance from the white county agent. Once again, the film rhetorically argues a message of white superiority, with the black agent being auxiliary to the white agent and needed only after the initial work has been accomplished. In actuality, however, these types of improvements were financed locally by the farmers, landlords, and civic groups and accomplished through the hard work of the community in cooperation with agricultural outreach personnel. Improvements often were accomplished via home demonstrations in which black instructors brought the community together in order to accomplish specific improvements on the property of willing farmers or landlords. As Thomas M. Campbell, the first black extension agent in Alabama, explains, these were often the goal of Tuskegee's Jesup Wagon (a rolling farm demonstration wagon) and later its Movable School projects, which allowed needed demonstration equipment to be carried into the field for the accomplishment of these projects.[16] Projects such as these preceded the federal county extension agency and were in part a model for the later USDA service.[17]

Although the film's intertitle states that the black agent becomes "necessary" to "assist the white agent," historically, this was simply not the case in Alabama. Black outreach was instigated by Booker T. Washington at

Tuskegee and sponsored by grants that he obtained; it was implemented with the expert supervision of Tuskegee's agricultural scientist, George Washington Carver, and carried out by black instructors, often in Macon County and the surrounding areas. By downplaying the central role of black extension agents, the film communicates the dominant belief that whites were the source of the ideas and the means by which the improvements were implemented. By continuing this theme, the film systematically empowers the white agents and rhetorically dismisses the importance of the black agents, instructors, and farmers—shamefully so, given the importance of the black agents' work and the meaningful role they played in helping the USDA to accomplish its goals in the southern states.

Tuskegee Institute

At this point the film switches to a sequence that highlights explicitly the Tuskegee Institute campus, specifically identifying it as a school for African Americans. Several scenes appear to take place on the school's campus, and intertitles laud the successes of the school with remarks such as "Tuskegee Institute, where thousands of Negro boys and girls have received practical training in trades and industries" and "It now has more than 114 buildings and 2,100 acres of land devoted to the farm and campus." These scenes introduce Tuskegee to the audience. Of course, Alabama's black farmers needed no such introduction. Beginning in 1892, Booker T. Washington had held an annual Negro Conference, where hundreds of black farmers learned ways to improve their homes and farms.[18] So this section of the film is designed to introduce and establish the credibility of Tuskegee to a larger audience who are unfamiliar with its work while also explaining that it is an institution for black education.

Having established Tuskegee as the center for black agricultural instruction in Alabama, the plot dramatizes historical reenactments showing that the Department of Agriculture sent a white representative to Tuskegee Institute in 1906 to confer with Booker T. Washington about "employing negro extension agents to work with their own people." The filmic version again makes the white USDA official the source of black extension work. In actuality this visit was made by Seaman Knapp, whose title was USDA Special Agent for the Promotion of Agriculture in the South.[19] Although it is not explicitly clear in the film, the tone implies that using black agents was Knapp's idea. In fact, it was Booker T. Washington who requested that the federal government appoint a black agent and who invited Knapp to visit Tuskegee. In two visits Knapp became familiar with Tuskegee's outreach work (particularly the Jesup Wagon). Scholar E. W. Crosby explains that Knapp was "convinced that the Jesup Wagon had prepared the way for demonstration agents. . . . Knapp agreed to give black demonstration agents a two year trial."[20] As a result, in 1906 Secretary of Agriculture Wilson appointed Thomas M. Campbell "a Collaborator of the Bureau of Plant Industry of the Department of Agriculture."[21] However, in the film the white USDA representative goes to a Tuskegee student and says, "I want you to go for the US Department of Agriculture among the Negro farmers in this state, and get them to demonstrate on their farms improved practices you have learned at Tuskegee." In light of the historical facts, this switch is worth noting. Tuskegee, Washington, and Campbell are the central players in black demonstration work in Alabama both before and after Knapp's involvement. Certainly Knapp is an important figure, and although he initially resisted the idea of appointing black agents, he soon understood they were imperative to the success of the extension service. It is true that without Knapp's support, Tuskegee would have had a more difficult time in convincing the USDA of the importance of black extension agents. The film's choice of Knapp as the source of the idea once again situates the power to change the fortunes of southern blacks within the cultural hegemony of white government agents.

Black Extension Agents

The next two intertitles explain that "a few years later the first negro agent, now a supervisor, confers with the [white] state extension director." The black agent says, "Collins and his neighbors are getting results from the white county agent's advice. A local negro agent should be appointed to work with them." Obviously convinced, the white supervisor agrees and signs a document and hands it to the black supervisor. The audience can assume this is the black agent's authorization to help the white field agent by working with black farmers. In historical fact the black supervisor was Thomas Campbell (he was the USDA's first black agent and later became a supervisor). The film suggests that the need for a black agent was prompted by the success of the black farmers owing to the advice they received from a white field agent. Crosby reports that the number of black agents grew from "repeated requests from Washington and Campbell";[22] however, the reason for the increase in the number of black agents was not so they could assist the white agents but because they were already doing extension work among black farmers. In the filmic version of these events, the local white supervisor gives the black supervisor permission to allow a black agent to work with the black farmers because they have seen results from the work of the white county agent. This scene acknowledges the fact that black agents worked with the African American community but once again gives the impetus and authority to the white agents.

The scenes that follow show the black county agent as he teaches the farmers better farming methods. They also introduce black female home demonstration agents. First the film shows these young women being taught by white instructors, and then it shows the black demonstration agents in the community assisting black women with domestic work, just as the male agent helps the men with farming.

The Movable School

The remainder of the film centers on the black agent and the home demonstration aides conducting a Movable School for Negro Farmers. Collins agrees to allow the agent to hold the extension school at his farm. On the day of the Movable School event, many black farm families come to Collins's farm and take part in the lessons, which consist of demolition, building, painting, cleaning, planting, nursing, and other activities. As the intertitle explains, "The farmer supplies materials for improvements, the family and neighbors do the work." The students at the school whitewash the Collins home, build a new chicken house, terrace the farm, clean the house, make flypaper, learn home first-aid techniques, take all of the furniture out of the house and clean it in the yard, shine shoes, set a table, store sweet potatoes, grade the road, and build a new vented outhouse. In just one day the Collins farm is transformed. At the end of the day, two white USDA representatives address the large group of farm families that have assembled. The white representatives again reinforce white superiority as they assess the work and evaluate its quality. Clearly they are pleased, as is Rube Collins, who, as the intertitle reveals, "becomes a contented farm owner; thankful for the prosperity extension work has helped to bring him."

These historical Movable School events are the focus of Campbell's 1936 semiautobiographical book, *The Movable School Goes to the Negro*

Farmer.[23] Much of the work described by Campbell is presented in these scenes. However, the film chooses to end the Movable School sequence with a visit from two white USDA representatives and again suggests that white USDA agents or supervisors were a key feature of black extension work. However, historical evidence does not support this position. In fact, county agent scholar Gladys Baker asserts that "the Negro county agent is not generally considered to be administratively responsible to the white country agent,"[24] and Campbell's recollections do not suggest even a moderate level of involvement by white agents in the Movable School project. Thus, the film is misleading in its recurring references to the importance of white USDA personnel to the development and work of black demonstration work.

When the messages of the film are compared with the historical records, it is clear that those messages are inaccurate. Thus, the film's form and content function rhetorically to argue that white agents were the source and focus of black agricultural outreach in Alabama despite the fact that Tuskegee personnel were both the impetus and the facilitators of black extension work. In other words, despite understanding that the black agents worked independently of white agents to enact positive change in the community of African American farmers via the long-standing projects started and maintained at Tuskegee, the filmmakers cinematically locate the power, authority, and knowledge within the white authority figures.

The Use of Stereotypes in the Film

Beyond the rhetorical focus on empowering white agents over black agents, the film further communicates traditional racist stereotypes of African Americans.[25] The first stereotype used is blacks eating watermelons. When introducing the Collins family children, the movie first presents Obe, then Ca'line. The next intertitle is "etc.," and the scene that follows shows a young black girl, approximately five years old, sitting on porch stairs eating a huge slice of watermelon. She buries her face in the melon, takes a huge bite, and then looks into the camera and smiles. In two more scenes the film portrays young blacks devouring watermelon. During the Movable School sequences a group of ten young women, volleyball champions, celebrate by standing in a semicircle and voraciously eating large slices of watermelon. They must bend over while they eat to avoid soiling their clothes with the juicy melon. The intertitle reads "The fruits of victory," and the celebration is obviously staged for the camera. The third watermelon scene is also at the Movable School event. One of the festivities is a watermelon-eating contest in which two young boys stand with their hands behind their backs and race to finish eating extra-large slices of watermelon. This manner of showing African Americans eating watermelon with wild abandon was a pervasive racist stereotype of the time. Cinematically speaking, this stereotype was widespread—so common, in fact, that a 1890s commercial film was simply called *The Watermelon Eating Contest* and consisted of the supposedly hilarious antics of blacks eating watermelons.

This stereotype served to demean African Americans. Film historian Peter Noble explains that the "chicken and water-melon eater" is one of "the principal stereotypes of the Negro in the American mind."[26] Cinema scholar Donald Bogle says this stereotype is referred to as the coon. The coon stereotype has two manifestations: the pickaninny (or child coon) and the adult male coon stereotype, sometimes called Rastus. The pickaninny was a "little screwball creation whose eyes popped, whose hair stood on end with the least excitement, and whose antics were pleasant and diverting."[27] However, as Bogle further explains, "The coon developed into the most blatantly degrading of all black stereotypes. The pure coons emerged as no-account niggers, those unreliable, crazy, lazy, subhuman creations good for nothing more than eating watermelons, stealing chickens . . . or butchering the English language."[28] This stereotype can be found in many films, including *Pickaninnies Eating Watermelon* (1910–11) and *Rastus and Chicken* (1910–11). The images of watermelon-eating blacks in *Helping Negroes* shifts attention from the accomplishments of African Americans and encourages viewers to focus on demeaning racial stereotypes.

Another common stereotype is the black jester or black clown/entertainer.[29] According to Bogle, this was the most common stereotype in the films of the 1920s. The jester portrays blacks as lazy, preferring to dance and cavort about, often for the pleasure of their white masters, rather than perform meaningful work. In the early minutes of the USDA film, the Collins children dance about to their father's fiddle playing. This scene, which is reminiscent of the happy slaves dancing for their white owners in the film *Birth of a Nation* (1915), must have been important to the filmmakers because it is the longest scene to that point in the motion picture, lasting over a minute in screen time, and the first (and one of the few) to be composed of three different shots edited together.

This stereotype is also used in the Movable School sequence. In the midst of the amazing day of productive work, the intertitle states, "A passing fiddler nearly breaks up the school." In the scene following the intertitle, a white fiddler rests against a fence while several black men dance, clap, and jump about. Again this is similar to many scenes that appear in blatantly racist films of the time. Media scholar Jane Rhodes says these types of films mocked and derided blacks.[30] *Helping Negroes* banks on this tradition to suggest that even these industrious black farmers cannot resist the siren call of the fiddler; they must stop what they are doing and dance just as their jester counterparts in mainstream commercial films of the era did. The intertitle suggests that the urge is so uncontrollable that the black men risk breaking up the school that they traveled miles to attend for their own edification. According to Bogle, these depictions of African Americans as lazy clowns were "used to indicate [blacks'] satisfaction with the system and [their] place in it."[31]

These racist stereotypes existed long before film in other forms of popular culture, such as music, and the USDA film capitalizes on this history. In the final scene the now happy and prosperous Collins family enjoys the material fruits of their labor, a new phonograph. Since there is no on-screen sound, the visual first shows a close-up of the record playing and then superimposes the words of the song on the screen in a mock sheet-music style. The song is Stephen Foster's "Old Folks at Home," a song often performed in blackface minstrel shows of the period. The song is about how a roaming black man misses his happy life in the land of slavery. Perhaps the most telling line in the song is when the sad singer admits that he is "still longing for the old plantation." The selection of this song to close the film reinforces the racist message communicated throughout the film: that blacks were not only eager to accept white authority but also prosperous as a result. The movie thus aligns itself with the racial stereotypes of the times and supports the racist view of African Americans as happy and content in their circumstances despite the fact that racial segregation and Jim Crow denied African Americans their civil rights.

An International Study of American Roads

It appears that the USDA believed the stereotypes used in the film were appropriate in 1921 and years later. In 1932 the USDA released an approximately thirty-two-minute film titled *An International Study of American Roads*. The film's length, for the time, suggests that the subject was important to the department, and, perhaps more significantly, it is one of the first sound motion pictures that the MPS produced.[32] However, this information is less important to my analysis than the fact that *An International Study of American Roads* utilizes a brief segment of film from *Helping Negroes*. The form of the 1932 film is somewhat interesting in that it is built around a driving tour taken by the delegates to an international meeting about road construction that was held in Washington, DC. The tour, which covered several states east of the Mississippi River, allowed the delegates a firsthand view of American roads. The movie is part travelogue as it shows the sights that the delegates witnessed along the way as well as the roads. During the segment on the southeastern region of the United States, an intertitle states "Here and there in the South the delegates enjoy the melodies and dances of a musical and carefree race." At this point the scene from *Helping Negroes* in which the Collins family children dance about to their father's fiddle playing is inserted. The sequence as it appears in *An International Study of American Roads* suggests that neither the USDA senior administrators nor the MPS staff saw problems in referring to African Americans as a "carefree race" (despite the reality of racial segregation, racial violence, and economic despair) or including this racist footage from the 1921 film in their first sound film. The USDA leaders were aware of the serious problems faced by southern blacks, but the stereotypical representations in films they produced do not seem to offend those who made the films or those who approved the films before they were released for circulation. In fact there is no evidence to suggest that anyone with decision-making authority over the motion pictures made at the USDA was overly concerned about the racist portrayals of African Americans in *Helping Negroes*.

Racist Ideology and the White Audience

The film's communication of racist stereotypes placed it squarely in the dominant cinematic culture of the time. Yet overt racist stereotypes are not the most significant aspect of the film in ideological terms. The common stereotypes in film were well known by 1921, and if the function of the film was simply to deride African Americans, then the movie would not need the educational focus or historic dramatizations that constitute its bulk. In order to access the larger issues at stake, the film's ideological stance must be fully interrogated.

The USDA films enjoyed wide release, as they were shown at the USDA, at state colleges and universities, in large cities and small towns, in the field, and at other venues, such as fairs, often with an implied or explicitly stated educational agenda.[33] For example, in 1922 a showing of *Helping Negroes* occurred in Birmingham, Alabama. A newspaper article announced that the film would be shown free of charge and "should be of special interest to Auburn men [college students, teachers, and alumni]."[34] The article, suggesting an instructional experience, reported that the movie demonstrates "the work being done by the extension service among negro farmers, it shows typical negro farmers and their families before and after they had the benefit of the extension work."[35] The overt connection between Auburn, the state land-grant college, as the headquarters of white USDA extension work in Alabama and the reference to extension work portrayed in the movie was clear to Birmingham residents of the time. The article is interesting in that it explicitly refers to the USDA's work with "negro farmers" without mentioning the black agents, and it specifically states that the movie will be of interest to the white "Auburn men."

Furthermore, a document from 1922 in the records of the secretary of agriculture's archives reports excerpts from letters about *Helping Negroes* that were written to the USDA. The excerpts are all very positive. The letters indicate that the film was shown at Georgia State College, at Philadelphia's Commercial Museum, and at the Delaware County Tuberculosis Association in Chester, Pennsylvania. One excerpt is of special interest because it is written by T. M. Campbell himself. He remarks: "I write simply to let you know that the reel (*Helping the Negro*) is meeting with unprecedented success. It was shown in a very large theatre in Atlanta Thursday and Friday, three times a day, and I am sure this reel is moulding [*sic*] sentiment for extension work generally and especially for Negro people."[36] The comments in the other letters echo Campbell's opinion that the film was raising support for extension work. It is clear that white audiences, northern audiences, and city dwellers were seeing the film and being positively swayed.

The ideological function of the film becomes clearer once the audience is considered. Since it depicted black men and women employed by a government-funded agency working to promote the well-being of poor African Americans, the film had the potential to challenge the prevailing cultural hegemony of the time. Indeed, the black extension workers were very aware of racists who resented their work as a challenge to the status quo. Crosby explains that "white sentiment towards the work remained a crucial factor in the program's development and expansion. . . . [And] black farm agents needed . . . an understanding of the racial relationships. . . . White sentiment ranged from outright hostility to full support."[37] Baker argues that "the obstacles that confront the Negro county agent . . . are tremendous. He must show proper deference and humility to the white leaders and never offend the more ignorant politicians and other leaders in the county."[38] Because of institutionalized racism, the black extension agents lived and worked in a precarious environment. An important part of their charge was the improvement of living conditions for black farmers, but they relied on whites for the financial resources required to do their jobs, partly because black extension agents faced funding problems. Archivist Dwayne Cox points out that "black extension work was funded less generously than white. In 1914, $10,000 in Smith-Lever funds went for extension in Alabama: $600 to Tuskegee; $200 to Alabama A&M . . . and the remainder to API in Auburn."[39] Moreover, black agents were dependent on the explicit and implicit approval of local whites in order to carry out their work among blacks. Many whites did not support their activities for various reasons, including racial hatred; so the black extension agents had to be very careful in how they conducted their work to avoid agitating whites.

The USDA extension officials certainly knew this, as demonstrated by a memorandum from Bradford Knapp, chief of Southern Extension Work and son of Seaman Knapp, to the secretary of agriculture. Knapp writes that a southern black extension agent "is attempting to keep down the racial agitation which seems to be running all through the South."[40] Like the black agents, the filmmakers must have realized that they needed to ameliorate any implied threat to the white establishment or risk angering elements of the white audience.

Regardless of the filmmakers' specific intent, the film accomplished this feat in three ways. First, the depiction of common racist stereotypes made the black farmers seem comical and childlike and thus less threatening. This helps to explain the absence of other racist stereotypes in the film. For example, the counter to the happy slave stereotype is that of the threatening black male. *Helping Negroes* does not utilize the threatening black male stereotype because it did not want to portray black men, such as the county agents, as dangerous to the white population.

Secondly, the movie supported the doctrine of separate but equal as it existed in the South—separate but *un*equal. As author Jerrold M. Packard points out, "The reality of 'equality' meant that the Negro part of everything would be indisputably and often grotesquely inferior to its white counterpart."[41] The state land-grant university referenced in the film is not named. It is referred to several times but never by name,

Alabama Polytechnic Institute (API), or even Auburn. In comparison, Tuskegee Institute is mentioned by name several times. There are at least eight scenes of the Tuskegee Institute campus and several others that are clearly supposed to be on the school's campus. Intertitles laud the successes of the school with remarks such as, "Tuskegee Institute, where thousands of negro boys and girls have received practical training in trades and industries." With this much attention to Tuskegee's campus, why not identify Auburn's campus, where the white extension supervisor was housed? Juxtaposing Auburn's campus with Tuskegee's might imply they were equal in importance, a position that many whites would object to.

Moreover, no white county agent is ever shown working with black farmers. The white supervisor meets with the black agent in the white agent's office. The white administrators come and speak at the Movable School after the work is completed. And although female instructors teach the black female demonstration agents, they do not go to the Movable School site and work with poor black farm families. In this manner the separation of blacks and whites is visually maintained throughout the motion picture.

Third and most significantly, the power to create change and the knowledge to enact transformation is held by the white figures in this film as evidenced by the focus on white agents. The film clearly makes this point on several occasions by using the word *white* as a crucial modifier and by clearly showing white men and women in positions of power. When pests are discovered in the cotton field, Rube reports the situation to his white landlord. The Collins family is allowed to live on the white man's farm in return for Rube producing cotton to financially benefit the white landlord. Historian Conrad explains that "Southern tenancy, like slavery before the Civil War, was an institution 'peculiar' to the South. It was not only an economic but a social and political order, the origins of which lay clearly in slavery and the plantation system."[42] The tenant farmer system inherently favored the white landlord and abused the black tenant.

Film scholar P. L. Simpson suggests that the way a film encodes ideological messages may "deflect attention from other possible" views of reality.[43] In this film the rhetoric shifts the possible focus from blacks as agents of change to whites enacting change for blacks. Rube's landlord sends for the white county agent, who dispenses knowledge to the crowd about how to raise cotton despite the boll weevil. After this event comes the dramatic shift in Rube's fortunes. The intertitle explains, "Following the agent's advice and with their landlord's approval Collins and his fellow tenant farmers begin to diversify their farming and to grow food and feed crops." In other words, Rube himself does not have the power to improve the lives of his family members; that power lies with the white landlord and white agent.

Moreover, the film presents the whites' ability to change the life of the black tenant farmers as miraculous. With the help of the white agent, "better farming has given the Collins family a better home, and a local negro agent to assist the white agent has become necessary." The scene is visually striking: the Collins family no longer lives in two separate one-room shacks with no windows; they now live in a fairly large farmhouse consisting of several rooms. This film's intertitle also points out that not only did the white agent provide the way for the Collins family to improve their lives but that he is also responsible for the hiring of a black agent.

This rhetorical appeal is supported further by the film's scenes in which a white Department of Agriculture representative goes to meet with Booker T. Washington at Tuskegee Institute to employ "negro extension agents to work with their own people." In other words, white people have the ideas and the means to enact them. In essence the white people dispense knowledge and have the power to bring about change; the black people reap those benefits. Only whites have agency or the means of acting to make positive change—this despite the fact that Booker T. Washington led Tuskegee to national recognition in agricultural education and outreach before the USDA was involved.

The film ends by explicitly stating that with the help of the extension agents, Rube Collins becomes a contented farm owner. Yet there is no evidence that Rube ever breaks the shackles of the tenant farmer system to actually buy his own property. In fact, the possibility that a tenant farmer could escape sharecropping was practically impossible. Conrad relates that sharecroppers were forced to remain in tenancy through poverty and debt to the landlord, and "that was the way the planter wanted it. He could not afford to let his tenants become too independent or self-reliant because he would lose control over them and this would cut his margin of profit."[44] The statement that Rube becomes a farm owner is a heavy-handed rhetorical ploy suggesting that a white landlord no longer exploits Rube. Although certainly possible, the historical evidence suggests that this was not what usually became of black tenant farmers, even with USDA help. Crosby contends that the black extension program ultimately failed: "The basic problem of the black extension program was its impossible goal of black agricultural self-sufficiency. Only very radical reform directed by massive federal intervention could have realized this. Such intervention, even if desirable, was a political impossibility."[45]

Helping Negroes sits the fence (to use an agricultural cliché), and in doing so its rhetoric is designed to raise support for the black extension program without suggesting that the program was a threat to white superiority and southern segregation. The film perpetuated common racist stereotypes and downplayed the importance of Tuskegee Institute, Booker T. Washington, Thomas M. Campbell, and others who performed black agricultural outreach in Alabama. The film locates the agency and power for change in the white agents, supervisors, and landlords and suggests that it is *their* good work that allowed black agents to help black farmers. Thus, the film functions ideologically to reinforce white hegemony even as it argues in favor of helping blacks become better farmers because ultimately the Collins family remained tenant farmers, a condition that benefited their white landlord.

Success of *Helping Negroes to Become Better Farmers and Homemakers*

Helping Negroes successfully achieved the goal of supporting the USDA's extension work with African American farmers while avoiding the ire of white audiences and politicians. As Thomas Campbell reported to the USDA in 1922, the film met with "unprecedented success" because it was seen by large audiences who, he felt, were positively influenced toward "extension work generally, and especially for Negro people."[46] In his report Campbell succinctly declared the two most positive and important facets of the film's work for the USDA: the large audience and notable exhibition conditions and the rhetorical success of the motion picture.

Likewise, E. A. Williams, the state agent for black extension work in Georgia, wrote, "The President and Faculty of the Georgia State College turned out in a body to see the picture (*Negro Extension Work* [*sic*]) and the teachers and students of other schools came in large numbers. We had farmers and their families to drive fourteen and eighteen miles from the country to witness this picture. All who saw the picture expressed appreciation for the work which is being done by the Negro agents."[47] Williams, like Campbell, points to the large audiences and the positive reception of the film as signs of its success.

Because *Helping Negroes* was received so well, the USDA was interested in the film's wide distribution beyond the normal circulation. One exception to the usual way the film was distributed occurred in 1929, when the USDA agreed to make a copy of the film to lend to Sangli Industrial and Agriculture School, in Sangli, India.[48] Although the department routinely received requests for its films from governments and organizations around the world, it usually refused to make copies for loan because of the expense involved. Instead the USDA allowed the outside interests (including foreign governments) to purchase copies of its films.[49] However, in this case, the department not only produced a print and lent it to the agriculture school, but it also approved an extension of the loan in 1931.[50] The USDA seemed pleased with the film's results and continued to consider it viable for distribution as late as 1934.[51]

Interestingly, the USDA archives do not contain written evidence that the film was ever openly or seriously criticized for its racist stereotypes, either internally or externally, and it remained on the USDA's list of available films for lending for many years. One explanation could be that the USDA simply ignored negative criticism it received about the film. However, on the basis of my review of the archives, I am sure the USDA in general and the MPS staff in particular were very attentive to criticism and would have pulled the film from circulation if they had thought that continuing to lend it might cause controversy that could endanger MPS funding. (Earlier the service pulled a very popular short-subject film about bronco busting because of complaints from humane societies.[52]) Therefore, it is safe to say that the film was successful and achieved its obvious goals for the USDA.

Helping Negroes provided support to the USDA in another way. The USDA archives show that from as early as 1917 the department received materials about the need to study the conditions under which African Americans lived in the South. For example, one letter from a "prominent colored citizen of Richmond, Virginia,"[53] dated 1917, suggested "that a bureau be established under the supervision of the Agriculture Department, to be known as the Negro Department of Statistics, Resources, and Commodities," the purpose being to find ways to "induce the Negro man and woman who has left . . . the South" to return.[54] The letter was accompanied by a short newspaper clipping that stated that more than 300,000 African Americans had left the southern states in the past months "seeking employment in Northern 'war' plants."[55] The USDA's response was to suggest that the writer contact the Department of Labor because the USDA "is not connected in any way with any efforts to induce labor in the South to move into northern States."[56] The USDA was aware of the migration but apparently did not wish to take it on as project. In 1918 a USDA assistant secretary, Clarence S. Ousley, mentioned the "migration of negroes to the north" in a short letter to Thomas Campbell, but his point was to clarify a statement he had made earlier, not to suggest that the department had a specific policy or stance on the issue.[57]

Interestingly, the strongest statement made on the topic is in the 1920 USDA publication "Extension Work among Negroes, 1920," which devotes one paragraph to the "very serious problem" of African American farmers leaving the farms for cities and towns.[58] However, the concern is not just about the migration, but also the "closing of the war activities and retrenchment in numerous manufacturing plants has resulted in wage reduction and laborers being laid off in large numbers. Much uneasiness and actual idleness among those who recently left the farm for the city has resulted."[59] Sociologist Martha A. Myers maintains that whites were concerned about this northward migration because "whites defined the 'proper' place for blacks as being not simply in the South but on Southern farms, well supervised and uplifted by contact with paternalistic whites."[60] In this vein, the USDA pamphlet argues that African American "leaders have done what they could to bring back to the farm those who desired to return. They very properly recognize that . . . the average Negro . . . is better off there [on the farm] than in the crowded city."[61] It is obvious that the USDA was aware that many blacks and whites in the South were concerned about the migration for a variety of reasons and wanted the USDA to get involved in solving the problem. But the USDA responded in ways that suggest it did not want to be directly involved.

The archive shows that the USDA responded by highlighting its African American farmer extension work.[62] In this way *Helping Negroes* was a cinematic testament to that effort and gave the USDA an official response that avoided the direct questions related to the black migration. As discussed earlier, in the film *An International Study of American Roads*, the USDA chose to refer to southern African Americans as a "carefree race." This choice of wording is especially curious because the USDA archives suggest that the department was well aware that the lives of southern African Americans were anything but carefree. From 1919 alone the archives contain three documents the USDA received that speak strongly against the idea that southern blacks were carefree. The first is a letter, perhaps a press release, on the letterhead of the National Association for the Advancement of Colored People (NAACP) and signed by John R. Shillady, NAACP secretary. The letter calls for a congressional investigation of lynching and quotes Mississippi governor Theodore G. Bilbo, who said that he was unable to stop a mob that hanged and burned John Hartfield, an African American, on June 26, 1919.[63] The second document is a non-USDA publication titled "Four Open Letters from the University Commission on Race Questions to the College Men of the South," which includes sections on lynching, education, and migration and an argument for a new reconstruction.[64] The third is a publication titled "Lynching: A National Menace, the White South's Protest against Lynching," written by James E. Gregg, principal of the Hampton Institute in Virginia.[65] These publications demonstrate that the USDA was aware of the racial hatred and violence in the South and that the majority of southern African Americans of the time could not be considered "carefree."

Moreover, long before the movie was released in 1921, the black extension workers were helping the USDA with a problem related to racial discrimination in southern states. In December 1918 a letter to Bradford Knapp, chief of the Office of Extension Work in the South, from T. O. Walton, acting director of Texas extension work warns, "I have heard a number of expressions that convinced me that the returned negro soldier will likely prove a very great problem in many communities in the south. . . . As an illustration of such difficulties . . . two negro boys returned from one of the camps and were thoroughly beaten by a bunch of white boys because they thought the negro boys were not quite as humble as they should have been." He goes on to mention a riot of "negro soldiers" in Houston eighteen months earlier in which a black extension agent named F. L. Blackshear was able to "control that very delicate and difficult situation," and suggests that "if some plan could be worked out by which men like Blackshear and others of his race could present their views to returning negro soldiers, it would surely prevent some serious difficulties that are likely to occur unless some such step is taken."[66] The USDA's response was to take Walton's advice; an April 1919 letter from Knapp to the USDA secretary states that "the three negroes whom we appointed last January to deal with the negro problems in the South are doing fine work." He mentions a "characteristic letter" from Blackshear and argues that Blackshear "is attempting to keep down the racial agitation which seems to be running all through the South, by insisting that the negro extension workers do not discuss that question."[67] The letter from Blackshear to his Texas extension agent colleagues warns them not to agitate race matters. He writes that he has learned "that a political group in Texas has criticized the colored Government workers in the South, declaring they were only a political machine working in the interest of the North." He denies this is the case and points out that this criticism shows "that there are influential people who do not favor a special negro extension force; second, that we must be careful . . . not to give grounds for the charge that we are mixing other problems with the extension work." He insists "as the head agent of the negro extension work in Texas that we avoid mooted religious, political or racial problems." He refers to the "race problem" and suggests that only judicious statements should be made about the race problem and then with the goal "to make peace between the races." Although it is never explicitly stated in any of the documents, the race problem appears to be violence against blacks by whites. Blackshear admonishes the extension personnel to leave the public presentation of the African American view to editors and other leaders. He closes with, "It is often said that if you give black people positions under the Government they will be unable to hold themselves to the task for which they were appointed and will tend to break loose and overstep the bounds of authority, reason and common sense. I hope none of us are going to make these words true."[68] Blackshear's work, along with that of two other unnamed black extension agents, and his letter received praise from his Washington superiors; hence, it is evident that the USDA knew about white-on-black violence in the South and believed the best way to deal with it was to encourage its black agents to avoid raising the ire of the white population in all cases and by trying to make peace when possible. As I have shown in this chapter, *Helping Negroes* achieved this by supporting the separate-but-equal doctrine.

In the beginning of this chapter, I mentioned that there were at least three reasons why the USDA decided to make *Helping Negroes* in the early 1920s. The archival evidence shows that the motion picture was successful in terms of two of those reasons: first, the film does a good job of highlighting the work of the extension service, and secondly, the film explicitly shows that the USDA was trying to improve the lives of southern black farm families. The third reason I cite is that the making of the film coincided with the release of USDA Circular 190, "Extension Work among Negroes, 1920." As I mentioned earlier, that publication included a short section that states that "a very serious problem during recent years in many farming sections has been the migration of Negro farmers to the towns and cities."[69] Although *Helping Negroes* implicitly suggests that blacks should remain on the farm, it ignores mention of the migration completely. The USDA archives do not contain criticism of the film for omitting reference to the black migration, but perhaps this omission was noted, because the next time the USDA cinematically visited its extension work in the black community, with *The Negro Farmer,* it made a point of declaring that blacks were best suited to life on the southern farm and should remain engaged in agricultural work.

Chapter 3

The Negro Farmer

In 1938 the USDA revisited the topic of southern black farmers in the movie *The Negro Farmer.* This twenty-three-minute talking film was subtitled "Extension Work for Better Farming and Better Living," and, like its predecessor, it focused on how the USDA's extension program worked to improve the lives of African American farm families in the South. With the success of *Helping Negroes to Become Better Farmers and Homemakers,* the department had good reason to update this theme in a sound motion picture. Well aware of the fact that audiences greatly preferred motion pictures with sound, the department mothballed *Helping Negroes* and its other silent films.[1]

There was another reason for showing how the USDA was helping southern African American farm families. The historical documents at the USDA archives reveal that for years the USDA was assailed by individuals, organizations, and politicians to do something about the migration of southern black farm labor to the North. As historian Alferdteen Harrison explains, this black exodus, also known as the Great Migration, refers to the period between 1915 and 1960, when as many as five million southern blacks migrated to the northern industrial areas from southern agricultural states.[2] Historians cite several interrelated reasons for the migration, including legal segregation, racial oppression and violence, and the search for better jobs and lives. Author Jerrold Packard argues that maintaining white supremacy in the South meant that "almost every facet of black Americans' lives was entangled on a daily, never-ending basis in racial bias. The increasingly obvious answer to Southern blacks became escape."[3] Reactions to the migration were mixed in both black and white communities; however, historian James R. Grossman writes that "either immediately or eventually, civil authorities and whites dependent on black labor sought to halt the exodus through coercion, persuasion, or some combination of the two."[4] Jack Temple Kirby says that during the 1920s and 1930s white landlords became frustrated at the loss of cheap workers as black tenants became scarcer due to migration away from southern farms.[5] Packard, too, argues that "blacks had long been too important to Southern agriculture . . . and losing this commodity was the last thing most white Southern farmers wanted to have happen."[6] The USDA leadership's reply to questions about the exodus or accusations that the department was not doing enough to stem the migration was to point to the resources it dedicated to its black extension work, which focused on helping southern African American farm families while avoiding explicit references to the migration.[7] It is from this perspective that *Negro Farmer* was finished and began its distribution in 1938.

The explicitly stated focus of *Negro Farmer* is the USDA's Live at Home program, a program that encouraged small farmers to make the most of their available resources, and its goal of improving the lives of southern blacks. Given the sociocultural times in which it was conceived and produced, the film is USDA propaganda that addresses the pressure the USDA felt as a result of the black exodus from southern farms to northern cities. The film's rhetorical argument is that blacks should remain on the farm where they belong and can be happy by identifying with the farm lifestyle and reaping the financial benefits available to southern farmers.

Sound and *The Negro Farmer*

The Negro Farmer is in one sense an updating of *Helping Negroes* with sound. The USDA needed a substitute for the 1921 film because the earlier silent motion picture was very successful in explaining the USDA's extension work among African American farmers while not angering white politicians or audiences. Indeed, black USDA agents Thomas M. Campbell and E. A. Williams independently reported to the USDA headquarters in Washington, DC, that *Helping Negroes* was well received by southern audiences and enjoyed good distribution. Campbell noted that he believed the film was doing the good work of shaping "sentiment for extension work generally, and especially for Negro people."[8] Similarly, Williams reported, "We had farmers and their families to drive fourteen and eighteen miles from the country to witness this picture. All who saw the picture expressed appreciation for the work which is being done by the Negro agents."[9] While *Helping Negroes* admirably achieved its goals, the USDA knew the film needed to be updated for an audience that was keen for talking pictures. By 1934 updating was a logical next step in a move to sound films that began five years earlier.

The MPS began its interest in sound technologies in 1929, when it began "preliminary experiments in the production of sound pictures."[10] By 1931 the MPS released its first three sound pictures; it also saw a 15 percent decrease in lending shipments of its films primarily because "the demand for silent pictures [had] tended to abate since the advent of talking pictures."[11] Throughout the early 1930s the MPS worked diligently to research and test sound equipment for production and exhibition in order to make intelligent purchasing decisions and to develop new facilities in order to produce sound films, tasks that were complicated by the USDA's small motion picture budget. Regardless of the challenges associated with the switch to talking pictures, by August 1932 the film service had functioning sound-recording facilities and portable field equipment.[12] By the later 1930s it seemed time to revisit the USDA's work with southern African American farmers via a sound motion picture.

Form and Theme of *The Negro Farmer*

Negro Farmer is composed of scenes of southern black farmers going about their business and the work of black extension agents teaching better farming methods while the narration briefly explains the vignettes and offers commentary. In this way *Negro Farmer* follows the basic form of *Helping Negroes*, with the earlier silent film's intertitles being replaced with voice-over narration. The film was shot primarily in Alabama and Georgia, with the narration and music added in postproduction. Thomas M. Campbell is credited as a supervising specialist for the film along with two other agents. The film was directed and edited by Raymond Evans, who became chief of the MPS in 1927.[13] The crediting of Campbell and Evans distinguishes the film as an important production to the USDA.

Although the title of the film does not specifically link it to the southeastern United States, the film subtly establishes the connection to the South in its opening title sequence. The first title states that the movie is presented by the US Department of Agriculture and shows a map of the United States as its background, with each state delineated by alternating lighter and darker shades of gray. The background to the second title, which presents the name of the film—*The Negro Farmer, Extension Work for Better Farming and Better Living*—is a closer view of the southeastern section of the United States from Texas to the Carolinas. The third title, with the same background, proclaims "Negro Melodies by Tuskegee Institute Choir, Wm. L. Dawson, Director." At this point the camera zooms in closer on the state of Alabama, a darker-shaded state between two lighter-shaded states, and the scene fades into a long shot of southern pine trees. These three elements—the Alabama map, the reference to Tuskegee Institute, and the establishing shot of southern pines—are all associated with the coming events of the film. This is not surprising given that *Helping Negroes* was also closely associated with Alabama and specifically the extension work performed through Tuskegee Institute.[14]

The avowed theme of *Negro Farmer* is how the USDA's efforts improved the lives of southern black farm families, particularly through its Live at Home campaign. This improvement is plainly stated early in the movie when an elderly black woman is shown rocking on the porch of an unpainted farm cabin. White-haired and covered in a quilt, she is talking with a black man in overalls, who sits on the porch next to her chair. The white narrator, George G. Farrington, says, "Aunt Sally Smith, born in Africa and long past her 110th year when she died in 1937, lived to see the hard lot endured by her generation and that of her children in some measure bettered by this [USDA] campaign to help Negroes help themselves." The film thus offers Sally Smith as a visual testimonial that the USDA's extension service programs improved the hard life of southern African American farm families by attributing this opinion to Sally Smith through voice-over narration.

Once establishing the improvement of the lives of black farm families as its theme, the film is roughly divided into five overlapping topics: gardening, crop and livestock improvement, sanitation and home improvements, 4-H club work,[15] and, finally, a segment on Booker T. Washington that focuses on an argument that blacks should be farmers and remain on the southern farm. All five subdivisions address extension work among southern black families while also simultaneously implying that blacks should remain engaged in farming. Consequently, in *Negro Farmer* the USDA's work to help improve the lives of southern black farm families is inextricably tied to the supposition that southern African Americans are best suited to life on the farm rather than urban life. The film ends with a Booker T. Washington quote in agreement with this supposition about blacks: "They are at their best living in the country engaged in agricultural pursuits."[16] The manifest theme of helping black families improve their lives by taking advantage of the USDA's Live at Home program is secondary to the latent goal of convincing the audience that southern blacks should remain on southern farms rather than migrate to the northern industrialized cities.

Constructing Reality in *The Negro Farmer*

The film explains that a major feature of extension work is a desire to improve the lives of southern black farm families. A large portion of the film shows how using the ideas provided by the extension service can improve the daily lives of the farmers by increasing production and by providing sources of pride in farm living. The cinematic result demonstrates that the farm can be more appealing and profitable; therefore, blacks will be happier and more satisfied on the farm. By highlighting its extension work among southern blacks, the USDA suggests to the audience that it is bettering the homes and farms of southern blacks and thus improving the quality of life of African American farmers and encouraging them to remain on the farms, where many pundits believed southern blacks belonged.[17]

To establish that there is a need for helping black farm families, the movie begins with a narrator named Farrington voicing the seriousness of the situation over a sequence of scenes showing the lives of southern black farmers. With a scene of black field workers, Farrington begins the narration: "There are about nine million Negroes in our southern states, and the majority of them live on farms." The film cuts to a scene of black women and children picking cotton as Farrington continues: "While many of these farmers have achieved independence and perhaps prosperity, all too many barely make a living. In common with all southern farmers, they have suffered from a train of evils that have their roots in poor farming, particularly one-crop farming." Although the narrator does not explicitly name the crop, for many farmers the one-crop evil was cotton, and the visuals make this connection by matching Farrington's "train of evils" remark with black farmers picking cotton. By focusing on the problems of overfarming cotton, the USDA made a choice to ignore other social and political wrongs that contributed to the evils experienced by African Americans, such as slavery, poverty, disenfranchisement, racial oppression and violence, the exploitative tenant farming system, and Jim Crow racism. Granted this is a short film, and its stated spotlight is on the USDA trying to improve the lives of southern black farmers. The choice to offer no comment, however brief, on the other evils that southern blacks faced is worth noting, especially since the agricultural conditions of the South, well understood by the USDA, drove black farmers to grow cotton. As Grossman explains, "Black farmers raised mostly cotton because they were forced to by landowners and local merchants who demanded a cash crop. Given the credit structure and sharecropping system, nobody had much incentive to diversify or improve agricultural land."[18] Moreover, the omission of any reference to slavery, racism, segregation, or political disenfranchisement is a conscious decision by the filmmakers to avoid suggesting that the sociopolitical and cultural effects of institutional racism played a definitive role in the difficult lives led by southern black farmers. Thus begins the construction of reality in this film: the plight of southern African American farmers is fundamentally the result of their poor farming practices. The film never explicitly denies that there are other factors involved, but it also never considers other factors. By doing so, it establishes the problem as a farming problem and then offers solutions to what the film rhetorically considers a strictly agricultural difficulty.

Farrington pushes forward with the agricultural solution, stating that for more than thirty years the USDA extension workers have "labored to eradicate these evils" through a "nationwide extension service" following the farm demonstration work begun by Seaman A. Knapp "long ago."[19] This establishes that the USDA is dedicated to eliminating iniquities through extension demonstration work and underscores its dedication by addressing its resource commitment as the visuals of black farmers fade to a map of "Counties with Negro Extension Agents." This map of the southeastern United States from eastern Texas to North Carolina is dotted with marks indicating the counties where black extension agents were stationed. The narrator elucidates: "This map shows the status of this work among Negroes in 1937. There are over 225 agricultural agents and about 175 home demonstration agents now working . . . among southern Negroes."[20] The map graphic is designed to ensure the viewer that the USDA has dedicated significant resources to this problem.

After highlighting its dedicated resources, the USDA film qualifies its efforts by explaining that the extension work among southern blacks must be limited to fundamentals. As the audience views a series of windowless, unpainted ramshackle cabins, the narrator proclaims that "in the nature of things, extension work among the poor farmers living in housing like these, on small farms, often on unproductive soil, must deal with fundamentals." The narrator never lists the fundamentals but offers an example: "For these folks the hound dog is a useful farm animal," because hunting and fishing is "a necessary means for providing fresh meat." From a cinematic perspective, it is interesting that the filmmakers chose to revisit the unpainted country cabin as a testament to the hard life that African Americans faced in the rural South. The movie returns to this motif several times after establishing it; the motif is introduced by Aunt Sally Smith sitting on the porch of an unpainted cabin in farm country. Sally Smith's cabin together with several other shacks and cabins highlighted in *Negro Farmer* visually underscores the poor housing conditions under which many rural black farm families lived. Historian David Eugene Conrad describes the typical tenant farmer's cabin as a place with no plumbing and with walls and roofs that allowed the wind and rain to invade the house.[21] Hence, the movie establishes that the poor living conditions African American farmers endure are the result of problems that must be addressed by attacking the fundamental aspects of life on the farm, such as inadequate housing. Moreover, visually linking black farmers with poor housing rhetorically reinforces the belief that blacks need whites to help them change their fortunes for the better, an argument that implicitly supports the myth of white superiority.

At the very least, the inadequately constructed homes and other farm buildings of the black families become visual icons for the poor conditions faced by the farmers, thus suggesting they have no means for improving their living situations. Farrington explains in voice-over narration, "The problem of the Negro farmer, like the proverbial problem of lifting yourself by your own bootstrap, is he has little or no cash income." Having pinpointed the lack of cash income as the main difficulty, the narrator states that "under these conditions the farm agent's problem is making the most with small resources." The poor conditions and lack of available resources force the extension agents to center their efforts on free or exceedingly low-cost measures to improve the farmers' hard lives. The audience is left to understand that the Live at Home program's focus is on fundamentals.

Therefore, the filmmakers chose this point in the film to present their most visually striking example of black poverty as a testimony to the need for extension work among southern black farmers. As the scene fades into view, the audience witnesses a one-room, windowless, unpainted, rough-hewn cabin in a dirt clearing as Farrington says, "Here is a one-room cabin in the cotton country. The farm: three acres of thin, sour land." Here the USDA once again connects growing cotton with poor, nutrient-depleted soil. The scene cuts to the cabin door as Farrington speaks: "The household—mind you they all sleep together on pallets in this little cabin—twelve persons," as one by one the twelve adults and children file out the cabin door and line up, standing or sitting on benches in the dirt clearing. The family appears untidy, docile, and expressionless. The narrator continues, saying, "This primitive home, this big household, this small thin-soiled farm—these exemplify the conditions in question at their worst." The serious task extension agents face is epitomized in this poor family's plight as the scene fades. This example presents the audience with the bleak facts in the starkest manner.

Negro Farmer creates a reality that directs the audience to see southern black farmers living in dire circumstances because of their poor farming practices. The reality, the movie further argues, is that the USDA has improved the hard lot of southern blacks and is continuing to do so by focusing on the fundamentals of farm life while its work is hampered by a lack of financial resources in the black farm community. Yet this filmic reality not only ignores the long-standing sociocultural and political wrongs done to African Americans but also oversimplifies the problems of southern black farmers to a limited agricultural perspective. As Conrad asserts: "Southern tenancy was a vicious self-perpetuating system . . . the workers were little more than serfs, held to the land by debt, ignorance, poverty, and dependence on the landlord. That was the way the planter wanted it. He could not afford to let his tenants become too independent or self-reliant because he would lose control over them and this would cut his margin of profit."[22] Despite the constructed cinematic reality, the real purpose of the film is to suggest that southern blacks belong on the farm and can be prosperous as farmers, a point that is reinforced throughout the remainder of the motion picture.

Cinematic Arguments in Favor of Black Farm Life

As mentioned earlier, the movie is divided into roughly five segments that address aspects of the USDA work with black farmers. Each segment presents a topic and illustrates how the situation of the families can be improved. However, rather than focusing on how the USDA explicitly improves the situation, each segment quickly moves to showing how money-making benefits can be derived from farm work, thus abandoning the film's self-proclaimed focus on fundamentals and leaping quickly to commercial concerns that highlight agricultural profitability. In doing this, the film belies its promise to help the miserably poor farmers and reveals its argument that not only should blacks remain on the farm but that they should also take pride in their farm lives and profit from their agricultural efforts (an outcome that the reality of tenancy made practically impossible).

The Home Garden

After being introduced to the poor family of twelve living in a shack in the woods, the audience soon sees how quickly the situation can improve. The subsequent scene is much brighter and optimistic as the audience views a black woman working in a garden while the narrator announces, "Live at Home is the watchword of the extension program, and a good home garden is the first objective, for otherwise a proper standard of physical well-being cannot be attained." The healthy, active woman appears in stark contrast to the passive family of the earlier scene. The lush garden in which she labors is vastly different from the dirt clearing of the earlier farm cabin and provides a more appealing view of farm life. Likewise, Farrington's tone is more upbeat and positive: "Here's a well-planned home garden, large enough to supply the needs of a big family with something leftover for market." His reference to "a big family" recalls the twelve individuals sleeping on pallets in the one-room cabin of the previous segment, while the remark about surplus produce for market suggests that cash income can also be attained by home gardening. The two problems addressed in this single brief scene present a much more optimistic view of black farm life for the audience.

The gardening sequence continues with two more gardens, the first highlighting okra and the second eggplants. The okra scene shows a black woman working in a tall okra stand as the narrator explains that canning garden surpluses "may go a long way towards supplying the family living." The audience of the time understood Farrington's reference to canning to mean preserving cooked fruits, vegetables, and meats in jars with sealed caps. Canning surplus food for later consumption was a home and community activity to ensure year-round access to food at a time when refrigeration was not widely available to farmers. Of course, having a surplus is a requirement of being able to can food for future consumption; so the USDA implicitly argues that hard work in the garden (in addition to proper care of the field crops) will produce a surplus to help with family needs. Thus, the film has moved from observing the poor family in the dirt clearing without basic necessities to showing how a good family garden can increase the health and vitality of the family both immediately and long term. Later the film revisits food canning: "Home canning of food products is a fundamental feature of the Live at Home program. A well-filled pantry is not only the best insurance against sickness, but it saves money." The scene continues as women prepare food for canning: "Since money saved is money made, the women and girls . . . add to the farm family income and . . . safeguard the family health." A scene of a farm woman approaching a set of shelves with neat rows of canned foods prompts the narrator to say, "These well-filled shelves show what the energetic housewife can do toward providing the family living." Hence, canning food is explicitly linked to increasing family income.

The third garden scene opens with a woman working in a large eggplant field and expands to show several men and women hard at work as the narrator says, "Here is a farm devoted largely to growing eggplants. Thus the garden may be expanded to furnish most of the cash income of the farm." This is a situation in which a family produces eggplants not as part of the family garden but as a cash crop. The picture is a dramatic shift from the twelve family members standing austerely in the dirt clearing in front of their rundown cabin. The eggplant scene shows a flourishing crop and several men and women actively engaged in farming endeavors. The differences in these two scenes is visually striking and invites the audience to be witness to the improved circumstances. The USDA, however, realized that jumping from a home garden with excess okra to a field of eggplants being grown as a major cash crop was a bit of a stretch; so the narrator qualifies the scene by adding, "In any event, the garden remains a major feature of the extension program." The leap from the desperate family in the dirt clearing to the prosperous group working on a cash crop of eggplants is accomplished visually with no concern for the lack of explanation about how a family would move from squalor to excess by way of the home garden.

Proclaiming that the garden is a "major feature of the extension program," the home garden sequence ends by bringing attention back to the extension service efforts within the black farming community, even though the sequence never shows extension workers or tells how extension workers help farm families improve their gardens. The a priori filmic argument asserts that home gardens are essential to a family's well-being and can improve a family's difficult life; therefore the extension service must help farm families produce better home gardens. Of course many viewers of the time would have been aware of the USDA extension service's communications on the importance and improvement of home gardens. For example, during World War I the MPS produced what it considered "special propaganda" footage on "war gardens."[23] Yet the choice to omit any obvious scene of USDA extension staff working with the families to improve their gardens is notable and suggests that an explicit visual connection was not deemed necessary by the filmmakers. The rapid progression of the garden sequence from an exemplary home garden to a commercial eggplant enterprise in four short scenes is economical as the film moves from a twelve-member family on the brink of despair to a thriving commercial farm. The implied lesson is that, with a little hard work, the family garden can become a commercial success. The fact that the garden sequence jumps rapidly from the fundamental concept that a home garden helps with the family's nutritional needs to showing its commercial potential contradicts the stated focus on fundamentals, because a commercial farming endeavor requires considerably more resources than basic gardening.

Later the film carries this idea further when a thriving roadside market is shown with several men and women shopping. The sign above the market, Macon County Curb Market, suggests that the scene was shot in Macon County, Alabama, home of Tuskegee Institute. Farrington punctuates the market's commercial aspects by saying, "a successful cooperative curb market. Such activities as this represent the occasional commercial development that comes from the Live at Home movement. Some of these markets are quite profitable." The garden sequence and related scenes about home gardening and its benefits are less about how to start or improve a home garden and more about the commercial benefits of a home garden. In other words, the garden provides an opportunity for cash income, and there is certainly no doubt that such income would improve the lives of black farm families. Additionally, cash income from gardening is seen as a way for southern blacks to stay on the farm rather than leave for work in the cities. Yet Conrad avows that sharecroppers had little opportunity for developing commercial options in their home gardens, particularly since they had to work in the fields from dawn to dusk: "Cropper farming, requiring close supervision by the landlord and obedience and servility by the tenant, gave small reward to individual initiative and self-expression. Essentially, it was a form of debt peonage."[24]

Crop and Livestock Improvement

Having explained the home garden's potential, the film then turns its attention to the work of improving cash crops through farm demonstration work. The screen brightens to reveal a black extension agent instructing a group of black farmers about the accepted standards for seed corn selection. The narrator says that "crop improvement work may begin with a neighborhood demonstration by the county agricultural agent. The lessons learned are fundamental." The scene then cuts to a black farmer walking through a tall, healthy stand of corn as Farrington declares, "Proof of the value of this [demonstration] work is such a cornfield as this on the farm of a cooperator." Again fundamentals are mentioned in this scene and visually tied to an abundant result. The lush cornfield is enviable; standing in verdant contrast to the dirt clearing of the twelve-member farm family seen earlier, it encourages pride in the corn and a positive identification with the farming lifestyle.

The film continues to expand on this theme as the scene changes to another healthy field and the narrator makes another comparison: "or this field of corn and soy beans. Note the luxuriant growth of the soil improvement crop [soybeans] between the rows of corn." The reference to soil improvement through the use of soybeans, a crop that replenishes nutrients to soil depleted by overfarming, is an implied reference to the evil of one-crop farming made in the film's opening and to the thin, sour land of the twelve-member family's farm in the cotton country. Experienced farm families would have understood that this is an allusion to correcting the soil-depleting cotton farming practices of the past.

At this point the film makes a bold step and moves from corn, an agricultural staple with feed uses on the farm beyond its cash potential, to a purely commercial crop. In a wide-angle shot, African American farmers are presented harvesting tobacco leaves by hand as they walk toward the camera. "Harvesting tobacco," Farrington says, "on the land of a farmer who works in close cooperation with his county agent on the work that is being done to improve the methods of handling this crop." Tobacco was a good alternative cash crop to cotton in the southeastern United States, and those familiar with its income potential would have seen this farm as a probable money-making enterprise. The USDA capitalizes on this understanding by cutting to a scene of two young women stringing tobacco leaves onto thin wooden poles that will then be hung in the tobacco barns for curing. It was often the practice of the farmer to pay laborers a small amount per pole for this work. The narrator develops this idea: "This farmer gives each of his children a definite share in the proceeds of the crop. These little girls are working not only for the family but for themselves. There's hope and personal pride back of this dexterity." The scene further moves the focus away from fundamentals and toward a more prosperous view of farm life that includes children as active wage earners for their own economic interests. In this dramatic step the docile and expressionless children of the worn and weary cotton farm are replaced by active and intent young women who labor to earn their own cash while helping the family succeed economically with a cash tobacco crop.

After briefly visiting a peanut farm, the screen brightens to reveal a cotton field with workers vigorously picking cotton as the narrator tells viewers, "In certain districts the extension workers have pushed the production of long staple cotton. A Negro family picking Sea Island cotton grown from seed secured through the help of the county agricultural agent." The sight of the cotton field here seems odd at first because the farming problems associated with cotton as the dominant cash crop on southern farms was well known and was alluded to in the "train of evils" remark used in the film to establish the need for USDA programs for southern blacks. However, here in the crop-improvement sequence, the movie has strayed from its focus on fundamentals to fully tout the benefits of commercial cotton production. Despite the problems associated with cotton farming, the truth was that cotton was still a notable cash crop, and the USDA was attentive to its production. However, the focus on cotton suggests that the film abandoned its pretense of focusing on the fundamentals of the Live at Home program to push the commercial benefits of farming instead.

The subsequent shot reinforces this notion. Three mule-drawn wagons heaped with cotton bales driven by young black men approach a cotton warehouse as Farrington describes the scene: "A group of 4-H Cotton Club boys each delivering his bale to the warehouse. The success of these boys has a good effect on the cotton farming of this neighborhood."[25] The USDA's commitment to improving cotton production is highlighted in the reference to the young men, who are members of the USDA's 4-H Cotton Club program. The next scene, which reinforces a link between the 4-H program and the cotton industry, shows a white man inspecting the various cotton bales for their quality and economic value while black men watch intently and the narrator tells viewers that "these bales are silent witness to the cause of better agriculture." This segment moves from the fundamental idea of using better seed corn to produce a healthier crop to showing how the USDA helped young black farmers to participate in the commercial cotton industry, with the latter example implying a significantly more important commercial business aspect than the former.

Like the gardening sequence, the crop-improvement sequence is less about the fundamentals of crop improvement and more about the commercial successes available to young black farmers who were willing to stay on the farm and join in the commercial cotton and tobacco industries of the South. Moreover, the reference to the 4-H club suggests that the young men were involved with an educational program that encouraged young people to identify with farming as a lifestyle and career. Their participation, the film suggests, was not only a silent witness to agricultural accomplishment but also an implied argument that leaving the farm for work in the cities was a poor choice.

The second major subtopic of this segment deals with farm livestock and begins with the significance of farm poultry. Voiced over a scene of a woman feeding chickens in her farmyard, Farrington extols the virtues of chickens: "The farm poultry flock is a factor of major importance . . . balancing the diet of the household as well as . . . [providing products] for sale or trade." In subsequent scenes of poultry-raising activities, the narrator links this practice to 4-H club members. First the narrator remarks that "poultry is a profitable demonstration for both boys and girls" while showing a large flock of chickens that are the "pride and joy" of a former 4-H club member; second, a young man is shown with his flock as the narrator explains, "This 4-H club boy took his chickens to school with him and with the proceeds for the sale of eggs is paying his way in a preparatory institution, a way to get an education if you have what it takes." The focus on fundamentals in terms of poultry farming is brushed quickly aside to focus on the cash potential and the use of cash to improve the opportunities available to this young African American in farm country. The scene also clearly states that poultry work is a source of pride and happiness for the farmer, thus encouraging a positive identification with farm living.

The poultry segment continues its focus on the business aspects of farm poultry by moving directly to a view of a commercial chicken enterprise as Farrington says, "Occasionally we find a man who goes into poultry in a commercial way. This is the plant of a farmer who specializes in White [Leghorns]." Thus, the poultry segment follows the pattern established by the earlier segments by introducing the topic as a fundamental of the home farm and then rapidly developing its commercial appeal.

The livestock sequence then moves to cattle as the audience witnesses a black county agent inspecting a milk cow while six young farmers watch intently. The narrator says, "Livestock improvement work is done largely through the boys and girls 4-H clubs; some Calf Club boys getting pointers on how to judge a dairy cow." This scene is followed by a shot of a row of six cows framed from the rear as Farrington quips, "Here is the business end of another 4-H Calf Club enterprise. These purebred Jerseys have a high record for butterfat production." Thus, the dairy cattle sequence continues the connection between the 4-H clubs as a way to encourage young men and women to be involved in the business of commercial farming.

In a brief sequence the film explains that "particularly in Alabama" veterinarians work with county agents to teach farmers how to "quickly and economically" vaccinate their swine against hog cholera before showing a small group of hogs "raised by a 4-H club boy" and proclaiming, "There's profit in well-bred swine." Later a county agent provides several onlookers with an outdoor demonstration of pork curing. The narrator describes the scene: "Home canning and curing of the family's meat supply saves cash. No outlay is needed for bringing home the bacon where farmers are thrifty and follow approved meat curing and smoking methods." Before the days of refrigeration a common technique for preserving pork for later consumption was the smokehouse method. Smokehouses were often small wooden structures where butchered pork cuts, such as hams, were hung and preserved through salt curing and smoking over an extended period of time. To highlight this method, the film shows a farmer entering a newly constructed concrete smokehouse as the narrator speaks to its importance: "A smokehouse is a necessity on the southern farm, for what is a home without a ham?" The scene shifts to a farmer stepping into his old wooden smokehouse to a voice-over: "But hams from the relatively inexpensive old-time farm smokehouse will smell just as sweet and stick just as close to the ribs as those from some more modern structure. The ham is what counts." These sequences combine to emphasize the importance of raising swine and smoking hams.

Ultimately, the livestock improvement sequence has very little to do with helping poor farm families improve their hard lot through fundamentals but is more concerned with showing how commercial livestock breeding is a business that provides profits through sales in local and regional markets. The film offers an appealing view of life on the farm as a commercial enterprise that can provide southern blacks with a ready source of cash income, an option that was completely out of reach for the twelve-member farm family held up as the most severe example of the need for the USDA programs in the black farm community. The movie does not address how that family is supposed to manage the transformation from malnutrition to making a profit raising pigs, especially since the exploitative tenant-farming system in the South encouraged landlords to keep their tenants at the "menial level" and avoid allowing them to improve themselves.[26] The narrator's earlier explanation that "the problem of the Negro farmer, like the proverbial problem of lifting yourself by your own bootstrap, is he has little or no cash income" is ignored in the livestock sequence, whose focus becomes how raising livestock commercially leads to financial success. The common references to 4-H activities suggest that active involvement in the 4-H clubs provides a link to the commercial benefits of farming for young people as well as a means for positive identification with farming as their future.

Sanitation and Home Improvements

The next segment of *Negro Farmer* opens with a group of three young women and three older women working at a table in a yard with several men and women watching. Farrington describes the scene: "A neighborhood dairy demonstration made by a girls' 4-H club under the direction of the home demonstration agent." The exhibit shows various dairy products as the narrator says, "Clean milk is a prime objective, and county home demonstration agents cooperate with the women in their communities in making all possible improvements in sanitation in the handling of dairy products." The dietary benefits of dairy are touted as Farrington explains that "in communities where such herds as this are the rule, a long step has been taken towards breaking the regime of fatback and sorghum under which pellagra has resulted and been such a curse." Pellagra is a disease resulting from a niacin deficiency that can lead to death. Its symptoms include skin lesions, mental confusion, and diarrhea. According to Conrad, the typical tenant family's "diet of cornbread, molasses, and sowbelly (fat salt pork), eaten three times a day, caused pellagra."[27] The USDA film uses pellagra as an example of how the nutritional improvement of the southern farm family's diet substantially improved the quality of life on the farm and links this to proper handling of diary products and sanitary practices.

The motion picture's references to fatback and sorghum are used to draw attention to the nutritionally deficient diet of poor southern farm families. Fatback is the layer of fat on a pig's back that is used as flavoring in southern dishes but is also eaten simply fried and salted. It is very inexpensive and therefore was quite popular among poor families in the South. Sorghum is a type of grass related to sugarcane and is used to produce sweet syrup similar to molasses. As it was plentiful and cheap in the southern United States, poor families and communities could fairly easily set up small outdoor homemade processing areas to produce sorghum syrup and thus provide themselves with a tasty treat, but one with little nutritional value in an already deficient diet. The reliance on cheap foods such as fatback and sorghum was a serious health threat for poor southern farm families; nutritionally deficient diets often led to malnutrition and disease.

To show how the USDA addressed these issues the film presents a brief montage of African American women working with various dairy products in their home as the narrator says, "An adequate family supply of dairy and poultry products combined with vegetables and fruits from the home garden helps to safeguard the family health." Following the pattern set in the crop and livestock improvement segment, the narrator then moves the topic to the commercial realm by adding, "Clean, wholesome farm butter; good homemade cheese; and other dairy products find a ready market when a surplus, in excess of family needs, is produced." Once again the focus quickly moves from the scourge of pellagra to producing excess dairy products for sale. Using this momentum, the movie goes immediately to a fully commercial endeavor by showing a large herd of diary cows in a fenced area near a barn as Farrington says, "Occasionally a farmer takes to dairying in a big way. Witness this big herd which furnishes Grade-A milk to a thriving southern city." The cattle sequence is brought to a close by mentioning how black extension agents have "inaugurated a better sires campaign" by breeding the small native cattle with "purebred bulls." The small local cattle are descendants of the original Spanish cattle brought to this area and are referred to as "scrub cattle" in the film; breeding them with purebred bulls greatly enhances their commercial value.

Despite establishing nutrition as the theme of the dairy/cattle sequence, the film rapidly abandons the serious nutritional needs of poor farm families and speeds quickly to the commercial benefits of cattle without making an attempt to explain how a family so poor that it must survive on fatback and sorghum could attain the means for commercial cattle farming, even on a small scale, without tangible material help from the government or private investment. The choice is clear: if the film's goal is to suggest that black farmers should stay on the farm, then the commercial benefits of farming must be highlighted in order to show the potential for income.

The film's rhetorical point is clarified later as it once again addresses sorghum, but this time the syrupy product has been divorced from its relationship to malnutrition and is presented as a good source of cash income. The movie presents a mule-powered mill grinding sorghum as Farrington says, "Sorghum mills, an institution of long standing, play an important part in the Live at Home campaign . . . [yielding] a staple that goes far toward reducing cash outlay for food; providing a cheap sweet that has its place . . . in the family diet." The narrator also notes that excess production can be sold: "Thus the Live at Home program always pays dividends; always in savings and sometimes in cash." So it is suggested that even sorghum is a reason to stay on the farm and enjoy the benefits of a cash income.

The movie then turns its attention to a shuttered window being opened from the inside of a rundown shack as the narrator proclaims, "Shuttered windows with no sash and no screen are the usual thing in the cabins of the poorer farmers. In cold weather these shutters are kept closed." The view switches to a young child sitting in the open doorway of a poorly constructed cabin. "Doors are seldom screened," cautions Farrington, "thus the first step toward providing better housing to guard against disease is to screen out the malarial mosquito and the typhoid fly."

The scene changes to a rickety outhouse with the door standing open. The narrator warns, "This is the common type of backhouse. Typhoid flies have easy access here." The scene quickly changes to a larger and better-constructed outhouse with a tin-sheet roof that the narrator endorses: "The approved type of farmstead toilet is not expensive but provides adequate protection from disease carried by flies." Although the second outhouse is new and better constructed, it is not clearly explained how the newer toilet provides more protection other than the closed door. The film does not elaborate further on the situation, although mosquitoes and flies were serious threats to health as Conrad explains that the lack of sanitation on tenant farms "led to malaria, typhoid, and many other diseases."[28]

Instead Farrington pushes on to a more palatable subject: "After sanitation come other home improvements that make for comfort and self-respect." The scene is an unpainted farmhouse in an unadorned clearing littered with farm debris. "Many a cabin . . . built in the simple and pleasing traditional style . . . have been made a home to be proud of by the application of a few cents' worth of lime whitewash or even pipe clay that costs nothing and the planting of a few native shrubs. . . . Such improvements take labor but little or no cash outlay, and the results are sometimes striking. Watch!" The scene transitions to the same cabin, but now it is whitewashed; the yard has been cleared of debris, a few plants have been added, and its picket fence is now also whitewashed. The narrator approves: "Now we have an attractive though unpretentious farmhouse, screened, whitewashed, and landscaped at little expense. Such improvements make for wholesome pride in the home place." The reference to the "wholesome pride" provided to the family is a subtle reminder that farm life can be a source of pride and enjoyment to southern blacks who remain on the farm.

The film extends the benefits of the attractive farmstead to the value of the farm crops as the scene widens to show a woman planting flowers in a grass lawn with a well-built and attractive craftsman-style home in the background as the narrator proclaims, "When the farm family takes pride in keeping the farmstead and lawn neat and attractive, we may reasonably conclude that the farm business is well managed . . . the field crops are likely to be good too." The dramatic change of venues from the rough-hewn ramshackle cabins in dirt clearings to this beautiful craftsman-style farmhouse set on a grassy hill with attractive flowers is a stunning visual shift that is related more to the home's architecture and grass lawn than to the addition of the flower garden. The climax of this section is therefore misleading, as the movie moves from a dirt-poor cabin to a well-constructed home with only the mention of screens, whitewash, and flowers. The cinematic result is certainly designed to be an appealing view of life on a southern farm that encourages the audience to envision an idyllic existence, but the visual payoff far exceeds the illustrated home improvement steps. The rhetorical thrust in this section is twofold: first, making the homestead neater in appearance affects the commercial viability of the farm in intangible ways, and, secondly, the more attractive farmhouse and farmyard is a source of pride for the farm family and therefore a better place to live and work.

The subsequent segment on plumbing likewise shows a grand jump in home enhancement without much explanation about how the improvements are accomplished by poor farmers. The segment begins with a scene of water being hand pumped from an exterior well as Farrington tells viewers, "In the South an outdoor bathroom is a possibility. We plug the drain pipe under the pump. The water is pumped directly into the bathtub and we are ready for our daily bath." The visuals do not reflect the commentary, however, since the pump is not located near a bathtub. The narrator says, "The modern pressure-tank outfit is better, and many farmers are installing standard plumbing," as the scene changes to a small indoor bathroom where tub, toilet, and sink can be seen in close proximity of one another. Again the rapid progression from an outdoor well water pump to indoor plumbing is dramatic, especially as the narrator reminds viewers that it is the policy of the extension program to make "the most of things within your means." The fact for southern tenant farmers was that even outhouses were a luxury.[29] The film does not explain how farmers who could not even afford screens for their doors and windows could install indoor plumbing, but neither does the home sanitation and improvement segment address the practical means for achieving such goals as a better-constructed home, a yard with grass and decorative flowers, or indoor plumbing; instead it relies on the visuals to show the preferred conditions by simply juxtaposing the inferior with the preferred homes and equipment. The movie leaves the final judgment understood without explicit argument: if the farm families enjoy the improved living conditions, then they will be happier and more content on the farm.

4-H Club

The efforts of the USDA's 4-H clubs are often highlighted in *Negro Farmer.* On several occasions the movie shows vignettes of demonstrations that feature black 4-H club members involved in improving life on the farm. For example, one scene shows a young woman making a presentation before a seated audience of men and women with several articles of clothing being displayed. Farrington explains: "This girl is a 4-H club state champion in dress making. All of the garments seen here are her handiwork. The knowledge and skill that these girls acquire in this work have a direct bearing on the family welfare." Later the film presents a similar display of garments at a county fair as the narrator informs viewers that the booth at the fair gives "the girl's 4-H club an opportunity to turn an honest penny through the sale of various articles to visitors." These sequences show how the 4-H club activities encourage young women to take pride in their activities and earn money.

The audience is also introduced to 4-H club camps when the film shows a large sign: 4-H Extension Camp. Keep Out. Members Only. The narrator explains that "the extension service sponsors 4-H summer camps in which short course instruction in agriculture and homemaking is given." A scene of several young black people filing past the camera in paper 4-H hats attests to the idea that the camp is well attended, and Farrington explains that campers come from across the state because the cost is low and there is "opportunity for study and profitable association." Subsequent scenes show male campers being instructed in judging dairy cattle and female campers being taught proper first aid practices. The 4-H club camp sequence ends with a group of young men and women playing and swimming in a lake as the narrator explains that there is also time for recreation. In this way the 4-H clubs suggest identification with the farm life, farming, and activities that can provide money-making opportunities.

Booker T. Washington

The final segment of the film focuses on Booker T. Washington and presents the last, and perhaps most influential, rhetorical argument for encouraging blacks to remain on southern farms. The segment begins with a scene of a truck with "The Booker T. Washington Agricultural School on Wheels" emblazoned on its side along with "Alabama Extension Service" and "US Department of Agriculture." The narrator says, "Last but not least among these extension activities are those of this Movable School operated jointly by the extension service and Tuskegee Institute to teach farming, homemaking, and hygiene"; following these words is a scene of a demonstration agent providing preventive treatment for infantile paralysis.

The film quickly moves from the practical demonstration efforts of the Movable School to focus on Washington's article "The Case of the Negro," published in 1899.[30] The scene fades into one where the camera looks up at Charles Keck's famous bronze statue of Washington, located at Tuskegee Institute. Gospel music swells up in the background as the Tuskegee Institute Choir sings "Down by the Riverside" with its easily recognized lyrics, "I ain't gonna study war no more." Over the background music Farrington proclaims, "Dr. Booker T. Washington taught that agriculture should be the fundamental pursuit of his race in America." The scene changes to a montage of black farmers working in fields as the narration continues: "And the extension service tries to help Negroes in their efforts to achieve objectives he so wisely laid down long ago." As the audience views black farmers picking cotton, Farrington continues: "'The Negro,' he said, 'must begin at the bottom and lay a sure foundation and not be lured by any temptation to rise on a false foundation. Progress by any other method will be but temporary and superficial. Since the bulk of our people already have a foundation in agriculture, they are at their best living in the country engaged in agricultural pursuits. . . . We shall prosper in proportion as we learn to dignify and glorify labor and put brains and skills into the common occupations of life.'" The song "Down by the Riverside" continues for several seconds until "The End" appears on the screen.

The film ends without mention of the Live at Home program or the improvement of the fundamentals of farm life that have been purported to be the core of the extension work among southern blacks. Instead, the focus is on the importance of southern blacks understanding that their place is on the farm and that any other type of lifestyle will result in failure. Although the entire film argues that blacks should stay on the farm, the ending employs Washington's words to underscore its meaning and intent. The gospel music provided by the Tuskegee Institute Choir adds religious gravity to the message. The choice of Washington's words is interesting in two respects. First, the opening paragraphs of "The Case of the Negro" state explicitly that black farmers should stay in the South and not move north; in Washington's words, "the Negro is at his best in the Southern states."[31] Second, the USDA chooses not to mention the other important points that Washington made in 1899; his essay also speaks against disenfranchising blacks by denying them their constitutional rights and suggests that blacks should not rely entirely on agriculture forever but should use agriculture as a foundation that can be built upon for other pursuits in the future. Given that *Negro Farmer* was made more than three decades after the publication of Washington's article and fifteen years after his death, it may be somewhat misleading for the USDA to focus entirely on his 1899 remarks concerning African Americans and agriculture while ignoring his other important points. After all, as Conrad contends, the tenant farming system in the South was unique because it was inspired by the plantation system and continued its racist socioeconomic and political hierarchy.[32] However, the rhetorical intent of the sequence is clear: the USDA film argues that blacks should remain on the southern farm engaged in agricultural pursuits.

The Use of Stereotypes in *The Negro Farmer*

Although *Negro Farmer* avoids the overt racist stereotypes used in *Helping Negroes*,[33] the film is plagued by the use of inferential racism. Stuart Hall explains that *inferential racism* refers to naturalized representations of race that have "racist premises and propositions inscribed in them as a set of *unquestioned assumptions*."[34] *Negro Farmer* avoids overtly racist images of African Americans such as the watermelon-eating pickaninny and the clownlike character dancing to fiddle music seen in *Helping Negroes*, but it contains scenes that uphold the racist ideology that maintains that blacks are docile and unintelligent. These include the scenes of the run-down one-room shacks, poor farms, and unsanitary living conditions. For example, the scene of the twelve-member family living in a one-room shack where they sleep on pallets in unsanitary conditions reinforces a view of poor southern blacks as passive and ignorant. Other scenes talk about unscreened, shuttered windows and doors and outhouses that breed disease. Yet the scenes neither discuss the debilitating sociocultural conditions that these families faced nor suggest that it is slavery and segregation that fundamentally created this state of affairs. As Conrad argues, the southern tenant farming system was an exploitative social, political, and economic order that kept tenant farmers in peonage with roots in the plantation system of slavery.[35] By using inferential racism, the movie perpetuates a traditional racist belief that blacks are too lazy to improve their living conditions without help from whites. One example of this perspective appears in the USDA archives, an item from a Tuscaloosa, Alabama, paper that states a belief that "wherever the Negroes are in the majority [in Alabama] there is stagnation and decay."[36]

As Grossman explains, many whites considered blacks "'unintelligent farmers' incapable of growing anything but cotton."[37] The film does show black farm families who are successful in one way or another, but only after the implied intervention of the USDA programs, thus continuing the theme that the power to enact change in the African American community is held by the white-controlled USDA.[38] By showing how whites can improve the impoverished lives of blacks, the film implicitly supports the doctrine of white superiority that was used to rationalize slavery, segregation, and other racist practices.[39] Finally, like *Helping Negroes, Negro Farmer* never shows blacks and whites working together, in continuing support of the doctrine of separate but equal established in the earlier motion picture.

Staying on the Southern Farm

Although the USDA's intent in *Negro Farmer* was to show how its extension work helped improve farming and homemaking for southern blacks, the form and content of the film reveals its rhetorical purpose of arguing that blacks should stay on the farm and be involved in agricultural pursuits. Most obviously, the film ends with Booker T. Washington's words: "They are at their best living in the country engaged in agricultural pursuits."[40] With this understanding, the commercial agricultural interests in scene after scene of the movie make sense. The film has little to offer in the way of suggesting how miserably poor farm families could pull themselves up by their bootstraps while it continually argues that agricultural pursuits were viable commercially and that southern blacks could take pride and joy in their farming endeavors. This stance is the result of the USDA's need to address its critics who thought it should do more to stem the black exodus while avoiding any overt mention of the migration by focusing on its black extension program.

The USDA was well aware of the northern migration of southern African Americans. The exodus of African American farm labor is mentioned in USDA archival documents as early as 1917.[41] In one instance the assistant to the USDA secretary responded to a letter of concern about the black migration by stating that the writer should contact the Department of Labor because the USDA "is not connected in any way with any efforts to induce labor in the South to move into northern States."[42] In 1920 the USDA publication "Extension Work among Negroes, 1920" devoted a short section subtitled "Leaving the Farms" to the "very serious problem" of the "migration of Negro farmers to the towns and cities. This problem has affected the whites as well as Negroes to a considerable degree." The pamphlet goes on to explain that the postwar closing of war-related plants resulted in the layoff of many laborers and that the "better class of Negro leaders . . . have done what they could to bring back to the farm those who desired to return. They very properly recognize that while conditions may not always be what they should be on the farm, the average Negro, dependent on his daily labor, is better off there than in the crowded city."[43] This early reference to blacks leaving the farm clearly states what will be the theme of *Negro Farmer* (eighteen years later), which also argues that African Americans are better off on southern farms.

African Americans across the country were concerned about the living conditions of southern blacks. In 1922 a letter from Silas Harris, president of the Negro National Educational Congress, to USDA Secretary Wallace refers to the large unemployment of "capable and deserving young Negroes." Harris believes that the USDA should employ these men and use them to generate interest "that would make the rural community more attractive" and "stimulate interest in agricultural life." Harris expresses his disappointment with the current government in Washington, DC, by saying, "I regret that President Harding has not been able to see his way clear to give, to my people, stronger evidences of this appreciation for their support, and his failure to do this is largely responsible for the present attitude of Negro citizens throughout the nation."[44] Harris, like Booker T. Washington, is arguing that these young African Americans would be better off on the farm and is chastising the Washington establishment for not doing more to help the rural black community. Harris's letter is a good example of how prominent blacks pressured the USDA to find ways to help African Americans to remain on the farm and to encourage the department to employ more blacks.

Similarly, H. L. Remmel wrote to USDA Secretary Wallace on Republican National Committee letterhead explaining that President Coolidge requested that he contact the secretary about his concerns related to the migration. Remmel writes that the loss of black labor in the South is "of vital importance to our planters and farmers, as the colored man and his family constitute practically the agricultural labor . . . of all of our southern states." Remmel suggests that the USDA form a commission of five African American men to travel broadly in the southern states to explain "the importance of diversified farming and to tell [southern blacks] that the south is the natural home of the colored man and his family."[45] Wallace's response reminds Remmel of the large number of black extension workers the USDA employed in the South.[46] By the early 1920s the USDA had its public answer to questions about its efforts to stem the migration; it simply reminded its critics of its wide-ranging extension work among southern blacks. The USDA argued that the black extension work, which helped southern black farmers improve their conditions, was the best way to address the migration.

Privately the USDA was interested in how blacks were faring in the North. In 1923 the secretary of the USDA authorized Thomas M. Campbell to travel to the North to investigate "what is happening to the colored people who have migrated."[47] In 1924 a USDA internal memo to the secretary regarding Campbell's trip reads: "It would appear that the negro who goes North is likely to stay there since he finds there larger opportunity for development, better wages, less restrictions, and better school facilities. Besides that he seems to make a good laborer in the industries, and many of the negroes have the ability to rise into the skilled labor class."[48] In response the secretary wrote "Interesting" at the bottom of the memo. So it appears that by 1924 the USDA chiefs were aware that migration was providing blacks some opportunities in the North, but despite this knowledge, *Negro Farmer* (made fourteen years later) continued to argue that the southern farm is naturally the best place for blacks.

The movie argues that the USDA was helping southern black farmers improve their lives even though it offered nothing of real substance for improving the hard lot of southern black farmers. Consequently, it would seem that the USDA supported the ideologically racist disenfranchisement of African Americans and the system that held southern blacks in a form of pseudo-slavery that Conrad called "debt peonage."[49] Packard explains that after Reconstruction, southern blacks were "ever more trapped in economic peonage."[50] For Conrad, southern tenancy was a "new plantation system . . . [where] planters maintained the necessary cheapness and docility of labor, and social and economic control."[51] Northern migration was not an absolute cure for problems that blacks faced in the South, but the USDA knew from its own investigation that blacks could find better conditions in the North. Moreover, as Grossman explains, many whites depended on black labor and wanted to stop the black migration.[52] The USDA's strong message that African Americans should stay on the southern farm ideologically supported the oppressive system that caused many of the problems that southern African Americans faced.

Chapter 4

Three Counties against Syphilis

The USDA produced a second film in 1938, titled *Three Counties against Syphilis*, which, like *Helping Negroes to Become Better Farmers and Homemakers* and *The Negro Farmer*, focused primarily on the southern rural African American population. Unlike the two previous films, however, *Three Counties* neither explicitly mentions southern black farmers in its title nor demonstrates the USDA's extension work with black farm families. Even the description of the film provided in the USDA film catalogs makes no mention of the black community.[1] Instead, the USDA produced this film for the Public Health Service (PHS) for the stated purpose of demonstrating the work of a mobile health clinic that moved around three counties in Georgia testing and treating the residents for syphilis. The intention of this public project was to eradicate the disease in the region. However, the movie did much more; *Three Counties* presented the PHS's medical view of syphilis in the rural African American population and provided insights into the government's reasons for syphilis testing and treatment there.

In this chapter, I demonstrate that *Three Counties* communicated that the PHS's program was fundamentally based in a racial medicine ideology that argued that blacks were a problem population; they were considered inferior to whites both morally and physically, and their perceived proneness to syphilis and other diseases posed a significant threat to white society. As historian James Jones writes, "To create support for health programs for blacks, health officers appealed to the self-interest of whites. . . . Segregation afforded no protection. Communicable diseases might be bred on the wrong side of the tracks, but blacks spread them into the white community daily."[2] *Three Counties* supported this racist perspective of black inferiority and argued that the government tested and treated the black community in order to protect white society from the threat of disease brought on by African American sexual promiscuity and physical and moral inferiority.

Significance of *Three Counties against Syphilis*

Three Counties is an important cinematic text in its own right, but its significance is increased by the fact that it was produced during the early years of the Tuskegee Syphilis Study, arguably the most notorious government-sponsored medical program in the United States, in which African American men endured "the longest nontherapeutic experiment in history."[3] Jones, author of the definitive book on this subject, *Bad Blood: The Tuskegee Syphilis Experiment,* explains that the US Public Health Service conducted "a study of the effects of untreated syphilis on black men in Macon County, Alabama, in and around the county seat of Tuskegee."[4] Given that the men were neither properly treated for syphilis nor appropriately informed of their participation in the program, civil rights attorney Fred Gray called the Tuskegee study "a misguided forty-year medical experiment."[5] The fact that *Three Counties* presents another southern PHS syphilis program for blacks that was conducted at the same time as the Tuskegee study increases the importance of this film to scholars because it cinematically presents the basic tenets of racial medicine as they were applied to syphilis testing and treatment in the 1930s and highlights aspects of the racist ideology that endorsed the Tuskegee Syphilis Study.

Moreover, the two studies are linked through an earlier endeavor funded by the Rosenwald Fund that was designed to detect syphilis in southern black populations. The Tuskegee study and the program that was carried out in the three Georgia counties grew from this earlier program, which studied syphilis in African Americans in six counties in the South. Dr. Thomas Parran, director of the PHS's Division of Venereal Disease when the Tuskegee study began, explained that the six locations chosen for the Rosenwald Fund study included "Macon County, Alabama, the most primitive of the communities studied and the most poverty ridden [and] . . . Brunswick, Georgia, and the turpentine forests back of it in Glynn County."[6] Although *Three Counties* does not mention the Tuskegee study, it is important to understand this film in terms of the Alabama study and the racial medicine ideology of the time concerning blacks and syphilis.

Three Counties is also an important USDA film because it takes place in southeastern Georgia (primarily in Camden, Glynn, and McIntosh counties near the cities of Savannah and Brunswick) and contains many scenes of local blacks working in the various industries as well as views of the lives of black men, women, and children in the late 1930s. Those depictions joined with the others in this book present a broad view of the representations of southern rural blacks in USDA motion pictures.

Establishing Credibility and Framing *Three Counties against Syphilis* in Georgia's Slave Past

The movie begins with a title bearing the seal of the US Public Health Service, declaring it a public health film with the endorsement of the US government. Likewise, a subsequent title informs viewers that L. E. Burney, MD, of the PHS was the film's medical supervisor. Like the PHS seal, this information offers governmental and medical professional authority and gravitas to the motion picture. The film's title sequence assures that it is a serious and medically informed presentation approved by the proper government agency and supervised by a physician.

The title sequence also locates the motion picture geographically in Georgia, much the same way that *Negro Farmer,* released the same year, established its location as Alabama. The title, *Three Counties against Syphilis,* appears in white letters over a map of Georgia and the lower section of South Carolina. The state of Georgia is identified in letters outlined in white, and Georgia is delineated from South Carolina with a darker shade of gray. The east coast of Georgia is clearly visible, and a nautical compass appears just off the cost of Georgia with the *N* (designating north) pointing toward the top of the screen. Thus, the motion picture establishes that its location is the southeastern state of Georgia.[7]

As the title sequence ends, the screen dissolves to a map, and the camera zooms to the coastal area of Georgia and three counties designated in light gray, in contrast to the rest of the state in darker shades. The map dissolves into a view of low waves rolling onto the Georgia coast from the Atlantic Ocean. The narrator, Alois Havrilla, begins, "South of Savannah the waves of the Atlantic break on the shores of the golden isles of Georgia. Once the rendezvous of buccaneers, later the scene of rich manorial plantations, these isles have character and charm." These opening remarks situate the location on the Georgia coast and evoke Georgia's slave-owning plantation past. The scene changes first to a view of Spanish moss–covered trees and then to the decaying walls of a two-story white building, roofless and gutted. The narrator asks viewers to "witness the picturesque ruins of the old slave hospital where Negro slaves were treated more than a century ago," further strengthening the connection to Georgia's slave history and explicitly referring to the medical treatment of African Americans in this area's past. Thus, the film establishes credibility while also linking the narrative to Georgia's slave past.

Racial Medicine versus the Generalizable Theme of *Three Counties against Syphilis*

The existence of a separate slave hospital is not surprising given the broad understanding that slaves were kept separate from their white owners in most institutional settings. More significant is the film's explicit reference to the fact that blacks were treated separately from whites. This separation in medical treatment long influenced medical practice, because many physicians considered blacks and whites different, an attitude that was the foundation of racial medicine. Historian Todd L. Savitt argues that racial medicine had roots in slavery with "political as well as medical ramifications. . . . [As] a part of a proslavery argument . . . to illustrate the inferiority of blacks to whites, to rationalize the use of the 'less fit' racial group as slaves . . . and to prove . . . that they [whites] recognized blacks' special medical weaknesses and took these failings into account when providing for their human chattel."[8] The reference to the old slave hospital situates the narrative squarely in the larger context of racial medicine and its racist ideology of the inferiority of blacks.

However, the filmmakers do not continue to focus on Georgia's slave past but quickly move to a brief homage to the area's connection with the poet Sydney Lanier and introduce the viewers to the port town of Brunswick, Georgia (the county seat of Glynn County); "important port of the old south [and] thriving seafood and industrial center of the new," explains the narrator. A series of scenes of men and women working in various industries—a cannery, a wood pulp mill, a turpentine distillery, and a cardboard box factory—are shown in quick succession as the narrator describes the area's industry and moves from the past history of slavery to a progressive view of Brunswick as a modern industrial city. The movie then transports viewers to rural Glynn County. As views of the port and other commercial concerns fade, the narrator informs viewers that "Brunswick [is the] center of a rural area [that] is typical of communities throughout America. Here a demonstration of syphilis control can be carried on, a test program for work in other places south or north." In this important scene the motion picture defines Georgia's syphilis testing and treatment program in these three counties as being generalizable to other communities regardless of their location.

While the film recognizes no other communities, areas, or states where the PHS had syphilis programs, just a few hundred miles to the west, in Alabama, the PHS was six years into the Tuskegee Syphilis Study. *Three Counties* suggests that Glynn County is the perfect area for the study because it is "typical of communities throughout America," but its focus, like that of the Tuskegee study, settles on the poor, rural southern black community. Within a few scenes, the film contradicts its claims of generalizability by concentrating on poor rural blacks. The focus on rural black communities is understandable within the framework of the racial attitudes that existed concerning syphilis. Historian James Jones points out that by the turn of the century, many physicians "depicted syphilis as the quintessential black disease" and long argued that black disease threatened white health.[9] So the selection of the black rural communities in Glynn and the surrounding counties is consistent with the perspective that primarily defined syphilis as a black disease and the need to control it as important to white health.

Brunswick and Glynn counties were not chosen for their generalizability. The PHS was well aware of the syphilis problem in the black community in that part of Georgia. In his 1937 book, *Shadow on the Land: Syphilis, the White Man's Burden*, Dr. Thomas Parran writes, "Brunswick, Georgia, and the turpentine forests back of it in Glynn County" were the locations of one of the six programs, in five southern states, financed in 1929 by the Rosenwald Fund to study syphilis and its treatment in African Americans.[10] The other counties were Scott, Mississippi; Albemarle, Virginia; Tipton, Tennessee; Pitt, North Carolina; and Macon, Alabama, home to Tuskegee. The syphilis rate among blacks in Glynn County was well known to the PHS when it planned this new treatment study in Camden, McIntosh, and Glynn counties in Georgia, and that knowledge played an important role in choosing this area. It is clear that this choice was not based on the generalizability of the program to any community in the United States; instead it was based on the knowledge that the black community had a high incidence of syphilis.

However, *Three Counties* overtly states that theoretically the program could be used as a model for any US community, suggesting that race was not a determining factor in the Georgia study. The transition from the ruins of the slave hospital to the area's modern industries and Brunswick's suitability as the site for a model syphilis demonstration is presented without explicit reference to the area's black communities. The scenes of mossy trees, harbor views, and local industries carry the viewer to a close-up of a white woman's hands, presumably a nurse's, drawing blood from a black man's arm as narrator Havrilla says, "The United States Public Health Service joined with the health departments of Georgia and Glynn County to set up such a project. This is not the story of syphilis the disease. It is the story of the public health organizations who stamp out syphilis." The motion picture quickly narrows its scope to a very specific focus on syphilis in southern blacks. The plot of *Three Counties* concerns the attempt by the PHS to rid Georgia of syphilis and to demonstrate how the three-county project would be suitable as a model for any community in the United States. However, the movie's content and form belie its generalizable plot elements and illuminate its focus on the black community and its reliance on the racist assumptions of racial medicine ideology.

Constructing Reality

Three Counties relies on the public's belief that it presents and comments on actual events and hence represents an objective reality. Thus, the film's form and content are constructed by the filmmakers to present the syphilis program and its patients to the audience in such a way as to support its claims. However, the historical evidence demonstrates that the film does not accurately represent the ideological reasons for testing and treating syphilis in rural southern black communities. The film is divided into seven major sections: (1) the weapons of medicine and public health, (2) finding and treating syphilis, (3) the black community, (4) convincing blacks to be tested and treated, (5) the bad blood, (6) the black church, and (7) the religious conclusion. In the following pages I address each of these segments and demonstrate how each presents part of the filmic reality of the syphilis program. I also explicate how historical understanding denies the constructed filmic reality in favor of a view of the program as a result of the racist beliefs of racial medicine.

The Weapons of Medicine and Public Health

An early focus of *Three Counties* is the science and medicine involved in diagnoses and treatment of syphilis. In order to visually represent medical science, the movie provides scenes of the procedures and equipment used to test and treat the disease. However, the reality of testing and treating syphilis in the late 1930s was grim. Once syphilis infection was confirmed, the treatments could last as long as seventy weeks and required constant follow-up to be successful. At the time, penicillin was not an option for treatment; instead, "in the early 1930s treatment consisted of mercury and two arsenic compounds called arsphenamine and neoarsphenamine. . . . The drugs were highly toxic and often produced serious and occasionally fatal reactions in patients. The treatment was painful and usually required more than a year to complete."[11] The film avoids the ugly reality of syphilis treatment and its possible side effects in favor of a progressive and scientific view of the treatment program. For example, the mobile clinic is introduced to the audience as the narrator enthusiastically proclaims, "Doctors on wheels! The clinical methods of the finest hospitals reach out to serve rural lives. Inside the [mobile clinic] trailer is a completely equipped clinic, electric sterilizers, refrigerator, fans, and lights. . . . No time is wasted in setting up work. Three minutes after arrival, treatment can begin." As the audience views patients having blood drawn and receiving injections, the narrator explains that the front section of the trailer is used for taking blood for laboratory tests and for injecting the syphilis treatment: "Arsphenamine injected into the blood kills the germ of syphilis, and the patient can no longer transmit the disease to others." The rear of the trailer is used for privacy: "Behind the curtain in the rear of the trailer is a complete examination room with its own equipment and sterilizer. Here bismuth is administered. Bismuth prevents the tragic relapses of heart and brain." These relapses are not elaborated on in the film; however, Allan M. Brandt, professor of medicine and science history, points out that the medical community at the time understood that "untreated syphilis could lead to cardiovascular disease, insanity, and premature death."[12] *Three Counties* avoids most of the unpleasant details of treatment, representing the work as modern, safe, and beneficial by demonstrating the equipment and set-up of the mobile clinic and making no reference to the fact that "abbreviated treatment might leave the patient asymptomatic but contagious, an outcome that others noted as particularly likely with uneducated and indigent patients."[13]

From the mobile clinic the film then moves to the scientific aspects of syphilis testing as the scene changes to the health department headquarters in Brunswick, Georgia, and a branch laboratory in Waycross, about sixty miles to the west. As the audience observes a technician working with a microscope, the narrator describes the testing procedures: "A little serum from the first [skin] sore is examined under the microscope. This is the earliest possible diagnosis for syphilis. The earlier treatment is begun, the surer the cure." The scene changes to a shot of the microscopic organisms viewed through the technician's microscope, several thin, black, wormlike organisms squirming in the fluid. The narrator tells viewers, "It would take thirty-five hundred of these syphilis germs laid end to end to make an inch."[14] The microscopic view of the disease adds visual interest to this medically focused segment of the film.[15] More significantly, the microscopic view of the syphilis germ adds scientific credibility to the motion picture and supports an objective scientific perspective on the disease and its treatment. This scientific perspective on the disease gives the film a means of detachment from the victims of the disease and encourages an impartial view of the testing.

Following the syphilis germ sequence, the audience sees a laboratory technician working with test tubes and pipettes: "Here blood tests are performed for the tricounty project. Tubes filled with blood specimens come from the trailer, from the Brunswick clinic, and private physicians." To emphasize the importance of the blood tests, narrator Havrilla explains that "trained laboratory technicians are essential; laboratories can't afford to make mistakes." The audience views close-ups of several test tubes being held over a magnifying mirror as the narrator dramatically proclaims the negative results for the first two samples and then a positive for syphilis for the third. While viewing the third sample, Havrilla says, "Note the heavy particles in the tube. That means syphilis." The narrator continues, "But one must be very sure. Let us double-check with the Wasserman test. This is another blood test for syphilis. . . . Let's see what it will show." A close-up of two pipettes holding fluid appears on the screen. The pipette on the left holds clear liquid; the one on the right holds a cloudy liquid. The narrator explains, "A milky fluid like the one on the right would mean syphilis. The clear tube is the check tube. Here are the tests; watch the tubes on the right." Again a series of close-ups appear as different samples are tested. As in the original scene, the first two samples are negative and the third is positive for syphilis. Havrilla explains, "Here our case of syphilis is verified. Even the layman can see the difference in these tubes." The motion picture makes a point of naming the Wasserman test and stressing that even untrained people (such as the audience members) can see that the results are unquestionably positive. These scenes, taken together, assure the audience that careful attention is paid to the credibility of the tests. However, in reality the Wasserman test's reliability was debated.[16]

Finding and Treating Syphilis

Unlike the Tuskegee Syphilis Study, in which known cases of syphilis were not properly treated, *Three Counties* was made as PHS propaganda designed to show that the organization effectively treated syphilis in Georgia. To that end, the film includes scenes of the testing process in order to create a version of reality. For example, in one scene a white man in a lab coat, presumably a doctor, is shown removing a box labeled neoarsphenamine from a medicine cabinet as the narrator, describing the treatment process, says, "In Brunswick the work resembles that of any city." A white doctor gives a white male patient an injection as the narrator continues, "Private doctors are given free drugs to treat all their cases of syphilis so that those who can afford to pay but little can go to their own doctors." Thus, the first scene of treatment is of a white doctor providing treatment to a white patient as part of the doctor's private practice. The narrator's words suggest that patients with scarce financial resources for medical treatment can afford this cure with the help of the PHS. However, the reality of treatment was that even when private physicians charged a fraction of their fee for their services, the number of treatments the syphilis therapy required made continued treatments "financially prohibitive."[17] Thus, the filmic portrayal of treatment affordability is not supported; the fact is that most poor rural southerners could not afford continued treatments, even at greatly reduced rates.

In keeping with the practice seen in the two previous USDA films under study, the filmmakers of *Three Counties* often avoid showing black and white patients together. The film does present white medical personnel with black patients and white program workers with black subjects, but it is always clear that the whites are in the positions of authority. For example, the scene of the white doctor and white patient moves to a scene of a black doctor injecting a black patient as the narrator underscores the wide availability of treatment to poor patients: "The private physician treats those who can't afford to pay at all." Matching the footage of the black doctor and black patient with the reference to patients who require free treatment is important; it suggests that even the poorest blacks can receive treatment and that the black private physician is willing to treat the patients for free. The film continues the free-treatment theme as the scene changes to a clinic setting and the narrator explains, "At the health department a modern public clinic treats two hundred seventy-five patients each week, free of charge." Hence, the film reiterates the point that even poor people are able to receive treatment.

The program's focus on testing in the black community is highlighted as the motion picture's view changes to a dusty school yard with dozens of African American children running and walking toward the camera. The voice-over explains that "systematic efforts have been made to find syphilis and bring it to treatment. Blood tests are part of school health examinations. Every child here has already been tested." The scene of the mostly middle-school-age children playing and the reference to their screening for syphilis are interesting for two reasons. First, public health practitioners knew that congenital syphilis was a problem and that some infants contracted the disease. More significantly, racial medicine had a long "preoccupation with black sexual behavior," and a well-perpetuated myth existed that "blacks matured physically at early ages and were more sexually active throughout their lives than whites."[18] Given the subtext of black promiscuity, the school yard scene underscores the belief that syphilis was prevalent in the black community. Thus, the film's comment that children were tested speaks to the fundamental racist belief that blacks were active sexually earlier than whites and that sexual promiscuity contributed to higher syphilis rates in the black community.

Additionally, with this scene all references to syphilis detection and treatment of whites stops, and the film focuses entirely on African Americans to its conclusion. Moving its attention solely to the African American population in effect presents syphilis as a black disease. Thus, the film exposes the PHS program as one intentionally targeting African Americans for testing and treatment.

The subsequent scene furthers a view that the PHS ensures that all blacks are reached regardless of their remote locations. The audience sees a small room where black men and women wait to be interviewed by a white woman sitting at a table. Havrilla begins to explain the significance of the scene: "In small communities in the outlying regions, the story is the same." He continues, "Here at Saint Simons, thirty-five patients come to the clinic each week," and the audience views a white doctor and nurse treating black patients. This scene of rural black community testing and treatment, working with the others in this segment, shows how the PHS worked to reach blacks even in small isolated communities.

In keeping with and highlighting this cinematic reality, a wooded area appears and the narrator dramatically announces, "But beyond the towns, beyond the doctors' offices, even beyond the roads are the woods." The onscreen image is of sawmills and turpentine stills as the narrator says, "This is rural land . . . fewer than ten people to the square mile. . . . How will one reach these people? . . . The health officer must reach these, too, if he is to meet his responsibility for the protection of the community." The problems of reaching all the infected are unambiguously declared for the audience: in order to protect public health, every black person, no matter how isolated, must be reached and treated for the explicitly stated "protection of the community."

The PHS method for testing in this remote area is the mobile clinic. These isolated individuals were people who worked in the forests of the area, a population that was well known to the architects of the earlier Rosenwald Fund syphilis studies in rural black communities. It became known during those studies that the turpentine forests were the remote home of syphilitic blacks. Hence, this sequence is the capstone of the motion picture's focus on assuring the audience that all blacks, no matter how isolated, received testing and treatment at the mobile clinic.

The Black Community

The motion picture shows that the modern, progressive way to travel into the piney woods to test and treat rural blacks is the mobile clinic. The film communicates that the white community was protected from black disease through this program. *Three Counties* introduces the mobile clinic's work in the black community by showing a car towing the mobile clinic, a vanlike trailer with a silver-colored top and windows all around it. The narrator tells viewers that the goal of reaching isolated blacks is achieved via the mobile clinic, which "travels throughout three counties: Camden, McIntosh, and Glynn." In order to represent the clinic's travels, the scene changes to a map of eastern Georgia showing the three coastal counties in detail, with roads (Camden is at the far south, McIntosh at the north, and Glynn in the middle). As Havrilla describes the clinic's trek, white lines show the trailer's path on the map. According to the cited statistics, in one week the trailer traveled 519 miles and treated 527 patients in 24 separate clinics. Next the audience sees a group of black men, women, and children standing in front of a small building and looking down the road as the mobile clinic drives into the frame. As the people board the trailer, the narrator says, "This silver top [trailer] is the symbol of the service. Along the road, the [mobile] clinic itself becomes a bus to bring patients who live in intermediate points to the treatment center." Later scenes show blacks coming to the clinic by oxcart and on foot (walking ten miles or more) and being delivered by school bus. Thus, the film's visual assurance that all blacks are treated and that the work is welcomed in the black community makes the program look even more effective.

The black community is first addressed through one of its most important institutions, the black church. The film shows the mobile clinic reaching its destination, where a group of well-dressed blacks have been lined up along the road waiting for its arrival. The narrator then proudly says, "At the clinic stop, the patients gather, next to church the biggest event of the week." The motion picture explicitly connects the syphilis program with church services, a theme that continues throughout the film. The connection to church service as an important weekly event to the black community is a significant rhetorical choice, since the church was considered the black communities' dominant institution.[19] According to historians C. Eric Lincoln and Lawrence H. Mamiya, "As the only stable and coherent institutional area to emerge from slavery, black churches were not only dominant in their communities but they also became the womb of black culture and a number of major social institutions."[20] By stating that the mobile clinic visit was the second most significant weekly occurrence in the black community, the motion picture argues that the program is important to the black community.

Also, by comparing the value of the weekly visits of the mobile clinic to black church attendance, the film attributes great significance to the testing and treatment of syphilis in the black community. The narrator unambiguously underscores the importance of the syphilis program when he again states that finding infected persons and bringing them to treatment is "the story of control. The treatment renders the patient noninfectious and stops the spread of the disease through the community." The filmmakers provide visual testimony to the importance of the clinic by showing three black men talking outside a country store as the narrator explains, "They gather at the rural forum, the country store, to compare notes on this new contribution to the life of the community." Repeatedly, the motion picture makes the connection between the mobile clinic and its success in the black community, overtly arguing that the goal of the program is the protection of black community life. Consequently, this segment constructs a cinematic reality in which the syphilis program is welcomed by African Americans as an important program to their community.

Convincing Blacks to Be Tested and Treated

After arguing the importance of the syphilis program to the black community, the motion picture turns its attention to the work of searching for new cases and following up with patients to make sure they are keeping up with treatments. In one scene the audience is presented with two black women sitting on the porch of a windowless country shack (much like the cabins seen in *Negro Farmer*). The women are questioned by a black woman with a clipboard. The narrator says, "The project is organized for finding cases and holding them in treatment. It takes seventy weeks to ensure a cure for syphilis. Even with free treatment it is not always easy to keep patients coming week after week." The scene changes to a white man talking with a black family on a porch of an unpainted country cabin as the narrator continues, "The [mobile clinic] trailer helps, but there are four follow-up workers, two white and two colored. The house-to-house canvass has proved a most effective means of getting blood specimens." These two sequences attest to the difficulties of testing and treatment that Dr. Parran explained in his 1937 book: "We realized that many of these people had never in their lives been treated by a doctor. Few of them had even seen a hypodermic needle. How, then, without the exercise of brute force could we get blood specimens for diagnosis?"[21] Parran mentions several strategies, including using as many blacks as possible in the program and pointing out that "in the first place, it is true in the South, by and large, that the Negro instinctively trusts the white man."[22] *Three Counties* was released one year after the publication of Dr. Parran's book and shows, in these brief vignettes, examples of both black and white public health workers contacting rural blacks for syphilis testing and treatment.

Continuing with this theme, the motion picture demonstrates how black midwives were trained to help with the testing. Havrilla describes the scene as the audience views a white doctor speaking to a group of black women in uniforms: "Midwives deliver four hundred babies a year in these counties. Well trained by the health department, they are licensed by the state. Here they learn about the blood test in actual demonstration. They carry forth the story and bring mothers in for prenatal blood tests." This approach is also in keeping with the advice in Parran's book, as he argues that blacks "trust the doctor—thanks to the fine character of many of our rural southern physicians. . . . [And] trusts the educated man and woman of his race."[23] Thus, the Brunswick program utilizes the techniques explained by Parran and used in the Rosenwald Fund study to gain the trust of southern blacks so that they could be tested and treated for syphilis without "the exercise of brute force."[24]

Three Counties also includes several scenes that show how the PHS advertised the program with flyers that specifically encouraged blacks to visit the mobile clinics. For example, one scene shows a black man wrapping an item in paper at a grocery counter as the narrator explains that the grocer includes the clinic announcement with every purchase. A close-up of the flyer reveals its content. The words "Colored People, Free Blood Tests, Free Treatments, Come and bring all your family" are followed by the place, the date, and the times that the clinic will operate. There is a scene, of flyers being posted at the country store as black men gather around; another scene shows a black woman nailing up a flyer at a sawmill as two black male workers watch. These scenes demonstrate that the flyers target African Americans and suggest that the black community is attentive to the program's promises as advertised by the flyers. The thrust of these scenes is that the flyers receive the attention of the targeted black community. The motion picture then demonstrates how the blacks were convinced that they needed to be tested and treated.

"The Bad Blood"

Three Counties suggests that the PHS program is welcomed because it addresses an important problem in the black community; however, when presenting the program to the black community, no reference to the word *syphilis* is made. Instead the program is advertised to African Americans as treatment for "the bad blood." Consequently, *Three Counties* is a cinematic testament to the PHS's use, and misuse, of the phrase *bad blood* to draw blacks to syphilis testing and treatment without properly informing them of the reason.

The narrator says, "Wherever the workers go they tell the story of the clinic to willing ears." To highlight this statement, the motion picture first presents a medium close-up of a black man's face that dissolves to a similar shot of a black woman's face as the narrator talks about "how bad blood makes people sick. How it destroys heart and brain." The scene dissolves to reveal four black children of various ages standing together as the narrator refers to "how it makes babies die." The scene continues with a series of dissolves to the faces of various black men and women, mentioning "how the blood test finds it and how the treatment, if taken long enough, is a certain cure." Given the title *Three Counties against Syphilis* and the fact that the film overtly references syphilis as the health concern many times, the connection between the euphemism "bad blood" and syphilis must be obvious to the audience. The change to using the phrase "bad blood" (after repeatedly referring to the disease as syphilis) occurs when the film relates how the PHS communicated the facts about the disease to the black community. The PHS used the phrase in order to encourage blacks to participate. This is important because in the black community "bad blood" did not explicitly refer to syphilis. Blacks understood "bad blood" to have a broad meaning that covered many ailments and was generally thought to be the cause of many symptoms from which blacks suffered. Thus, the use of the term played a vital role in convincing many blacks to seek testing and treatment for their ailments.[25] For example, Jones argues that Macon County residents responded favorably to the call for free treatment of "bad blood" because "blacks thought they were being tested and treated for whatever ailed them."[26] Moreover, many blacks did not know that syphilis was a venereal disease, and often patients were not properly informed that they were contagious and could spread the disease through sexual intercourse.[27] Therefore, the PHS benefited from the ambiguity of the term in the black community as it encouraged blacks to seek testing.

This is a significant point because it suggests that many blacks were afraid to submit to having a blood specimen taken; so finding ways to encourage all blacks to be tested was crucial. For example, the film presents two black men working in a pine forest. A black woman approaches them and begins taking a blood sample as Havrilla says, "The blood test serves a double purpose. First it is a dragnet for syphilis. Second it is an educational device." Havrilla addresses the educational function of the specimen-collecting process by explaining, "A man who has seen the blood test loses his fear. The man who hasn't had one wants it. Everybody talks about it. The blood test becomes a demonstration and a symbol." *Three Counties* shows that members of the African American community embrace and desire blood testing because they see and hear about their friends, loved ones, and neighbors receiving the test. Since "bad blood" was a general term used to refer to a host of ailments, the use of the term by the PHS suggests that its testing program addressed a multitude of possible health problems in the black community and that submitting to a blood test was encouraged broadly for whatever ailments the patient suffered.

The PHS evaded explicit references to syphilis and to "bad blood" as a descriptor for a venereal disease. In the film African Americans are shown at their homes, at school, shopping, and at places of work while always avoiding an overt association between "bad blood" and sexual intercourse. The motion picture breaks with this practice in a brief scene of young black couples dancing. Havrilla makes a remark during the scene: "Youth faces its problems more directly, honestly, and spontaneously than did its father." The scene dissolves to reveal a white man and woman taking blood specimens from the young African Americans at the dance as the narrator explains, "Here the blood test carries its message to a country dance hall." The scene itself is not odd for this film—the social environ of a dance hall is a completely understandable addition to the other locales presented—but the narrator's obscure remark is not overtly explained. One understanding of the comment is that older blacks were less likely to understand that the test was for syphilis than those who were younger. Regardless of the filmmakers' intent, this is the only scene in the film that associates heterosexual social interaction with syphilis.

The Black Church

Having dealt with scenes of school yards, turpentine forests, and a dance hall, *Three Counties* reserves its longest sequence for its attention to the black church. The sequence begins with a scene of a congregation from the minister's point of view and continues with a series of shot and reverse shots presenting the congregation as they listen attentively to the minister's message. Havrilla underscores the importance of this sequence by explicitly stating that "the greatest help [for the blood testing program] comes from the churches. Pastors are ever willing and glad to cooperate. They urge their people to join in the program." The motion picture abandons the secular nondiegetic musical score that accompanies many previous scenes in favor of a black choir softly humming a gospel hymn. The narrator explains that the pastors "tell [their congregations] of the disease which threatens health and life and the generation yet unborn. They urge that the message will be carried forth to the countryside that all may know and walk in the light of health." The "walk in the light" reference parallels the words of the hymn being softly sung by a choir on the sound track, "Walking in the Light of God," and so strengthens the relationship between the syphilis program and the black church. The scene changes from a focus on the African American minister and congregation to a white man and woman sitting at a table in the front of the church as the congregation looks on from the pews immediately in front of the table. The minister takes off his coat so that the white PHS workers can take his blood specimen during his sermon. The narrator explains that pastors "take the blood test as a demonstration and urge their parishioners to follow them." The rhetorical significance of this scene is great; the motion picture has moved from the weapons of medicine and science to the very heart of the African American community, the black church, to argue that the blood testing program is for the black community and has the full support of the church.

Lincoln and Mamiya explain that the black church was the dominant institution in black communities,[28] and the USDA capitalized on this by incorporating important aspects of the church into two of its motion pictures. For example, in *Negro Farmer* the USDA chose to highlight traditional gospel music in the sound track during the sequence dedicated to Booker T. Washington to provide religious significance to the USDA's argument that African Americans should not migrate from southern farms (see Chapter 3). Likewise, in *Three Counties,* a traditional gospel hymn is used in the sound track to add religious gravitas to the PHS's syphilis testing and treatment program in the black community. The narrator strengthens the church's commitment by stating that the pastors are happy to help encourage their congregations to participate in the syphilis program. Thus, *Three Counties* claims that the program has the explicit approval and support of the black church. These scenes link important elements of the church to the USDA's message by associating the hymns and the sermon to the cinematic argument. It is not required that these elements explicitly support the USDA efforts; what is important is that their association with the black church implies that the messages have a religious significance that is not inherent.

Concluding the Film on the Religious Theme

The motion picture moves from the black church segment to its concluding segment. Havrilla enthusiastically proclaims, "Such is the story of a community mobilized, a community which met syphilis with the weapons of medicine and modern public health. Whatever you may do in your community, in these three counties of southeastern Georgia the days of syphilis are numbered. Syphilis will be the next great plague to go!" Medically, there was no doubt that syphilis was a plague in the black community, but the use of the word *plague* also strengthens the religious association between the syphilis program and the black church because it recalls the biblical plagues of the Old Testament.

As the audience watches the African American congregation, the scene slowly dissolves to the mobile clinic car with its silver trailer driving toward the camera on a country road. The background music is the hymn "Walking in the Light of God," sung by the Hampton Institute Choir. This scene dissolves into a series of shots of various African Americans traveling toward the camera on foot. One scene shows African Americans riding in an oxcart, with the choir singing the gospel hymn and repeating the line "We are walking in the light" as the scene changes to the clinic trailer with patients waiting outside for testing and treatment. The choir continues the hymn as the film presents a white woman sitting at a desk looking at patient forms with treatments listed for each, and then the movie cuts to a view of the mobile clinic trailer driving away down the country road. In this way the movie abandons all of the science of earlier sequences to focus the audience entirely on the relationship between the program and the black community via the black church and the gospel hymn. The audience sees the black community embrace testing and treatment, literally walking miles to reach the clinic, while the hymn interweaves the PHS program with the significance of the black church, thus providing religious gravitas to the testing and treatment of syphilis in these rural areas. The concluding sequence ties the tricounty syphilis program exclusively to the black community. There is no doubt, at the end of this film, that the program was designed to test and treat African Americans. The cinematic reality communicated is that the program is embraced by the black community, endorsed by the black church, and framed in a religious context.

The Use of Stereotypes in the Film

Three Counties does not contain the overt racism communicated in *Helping Negroes*, but the film is nevertheless racist in form and content. As cultural studies scholar Stuart Hall explains, "The media—sometimes deliberately, sometimes unconsciously—define and construct the question of race in such a way as to reproduce the ideologies of racism."[29] Despite its lack of overt racial stereotypes (such as the Uncle Tom, mammy, happy slave, or devoted domestic characters) seen in many films from the 1920s and 1930s, the portrayal of blacks as helpless individuals who do not understand the need for testing and treatment of syphilis is stereotyping. Hall explains that the underlying message of many racist media representations of blacks is that they are "the dependent peoples, who couldn't manage for a day without the protection and know-how of their white masters."[30] In *Three Counties* whites are represented unfailingly as the knowledgeable purveyors of science and medicine while blacks are consistently shown to be the somewhat childlike patients in need of coaxing and cajoling in order to be tested and treated. Donald Bogle argues that the earliest racial stereotypes were designed to entertain the audience by stressing the inferiority of blacks.[31] Consequently, the portrayal of whites' superiority and blacks' guileless need for protection argue that African Americans are inferior. In this motion picture, individual overt racist stereotypes are downplayed in favor of an overarching exploitation of the belief in the inferiority of blacks.

The white medical personnel (doctors, nurses, lab technicians) are presented in positions of authority over their black patients. The separate-but-equal law of the time ensured that most whites were in positions of authority (both culturally and institutionally) over blacks in the South. The motion picture represents the African American community members as passive receivers of medical treatment who cannot understand the medical and scientific reasons for the PHS program and so must be enticed into treatment by whites or by black personnel and ministers working with the PHS. In the film most African Americans are virtually childlike in their unsophisticated understanding of the medical treatment and their trust in the white officials. Bogle points out that the "childlike lackey" stereotype enforced the view of blacks as inferior to whites.[32] *Three Counties* capitalizes on this pervasive stereotype to represent blacks as simple, hence inferior to whites. Even in the few scenes where blacks are presented as community leaders, for example, the minister in the church sequence at the end of the film, whites are superior. In the scene with the minister, white medical personnel draw his blood; he submits to their medical procedure in order to encourage members of his congregation to do likewise. The narrator explains that allowing the members of the congregation to watch the minister submit to the procedure helps the congregation members overcome their fears. In so doing, the movie suggests that African Americans must have their childlike fears assuaged and trust the superiority of white medical personnel.

Moreover, the narrator and narration assume a patronizing tone that underscores the representation of the African Americans as unable to fully understand the public health work being conducted. For example, the narrator mimics a black preacher explaining to a church congregation that they must all get tested for the "bad blood," thus relying on a generic term that covered a variety of ailments rather than explaining that the test was for a venereal disease. Reliance on this term and other generic statements, such as "the disease that threatens health" (when referring to communication with the black community), underscores the paternalistic stance taken by whites in their dealings with the black community.

The narrator also enthusiastically claims that "next to church, [the clinic is] the biggest event of the week" for the black communities, thus suggesting that being tested and treated for syphilis is an enjoyable event for the community that is universally regarded with positive anticipation. The community members are seen excitedly talking about the upcoming clinics, waiting passively for the mobile clinic to arrive, and traveling miles on foot and by oxcart in order to attend the clinics. Equating the weekly clinics with the importance of weekly church services argues rhetorically that this is a significant occasion for the community. Undoubtedly, these rural communities needed medical treatment, and the availability of a free mobile clinic drew community members to its services; however, the narrator's expression of the community's anticipation of the event suggests a naive view of the clinics by African Americans.

Even though *Three Counties* avoids specific racial stereotypes, its treatment of blacks as inferior to whites and existing in a state of childlike need of white protection is the very foundation of racist stereotypes in cinema. Despite the film's use of medical language and scientific laboratory demonstrations, the filmmakers chose to present a simplistic description of the program to the black community, a choice that defined the black community as culturally and intellectually inferior. Although the movie avoids blatantly mocking African Americans, its paternalistic stance toward the black community as unable to understand the medical causes and treatment of syphilis emphasizes the belief in the inferiority of blacks that existed since the earliest days of cinema.

Racial Medicine and the Problem of Race

Ultimately, *Three Counties* is about racial medicine. The movie's form and content present an argument based in a racist ideology that defines blacks as the source of problems in white society. Stuart Hall explains this as the "ideological construction of black people as a 'problem population.'"[33] *Three Counties* constructs a filmic reality of African Americans from a perspective that views blacks as a population that threatens the health of the white community and hence is a problem population. In this case the problem is the threat that syphilis presents to whites.

Three Counties creates a multilayered reality for the PHS syphilis testing and treatment program. Despite its early suggestion that the program is generalizable, the film quickly changes to a singular focus on the African American community by creating a strong connection between the PHS program and the black church as the dominant institution in the black community. Further, it argues that the weapons of modern medical science are brought to bear on the scourge of syphilis while resolutely clinging to the ambiguous and misleading practice of referring to the disease as the "bad blood" in the context of communicating to the black community. Ultimately, however, the motion picture produces a confusing narrative that abandons the early discussion of medical practices and the science of the mobile clinic to argue that the program uses effective means to treat syphilis in the black community. Understanding the inherent racist foundation of racial medicine explains this confusing narrative and elucidates the filmmakers' choice to rely on inferentially racist views to construct *Three Counties*.

Jones explains that the concept of racial medicine existed long before the 1930s as medicine and medical science suffered the effects of racism.[34] Racist attitudes affected the practice of medicine and influenced medical misconceptions about the health and medical treatment of blacks; thus, racial medicine played an important role in white perceptions of blacks and syphilis in the 1930s. According to Jones, "Advocates of racial medicine argued that differences in natural immunity, degree of susceptibility, and relative severity of reaction to various diseases often separated the races."[35] Medical and science historian Allan Brandt agrees and ties these beliefs to black sexuality, stating that "interest in racial differences centered on the sexual nature of blacks. The Negro, doctors explained, possessed an excessive sexual desire, which threatened the very foundations of white society."[36] Brandt says that some early-twentieth-century physicians believed that blacks were "especially prone to venereal diseases" because of "lust and immorality, unstable families, and reversion to barbaric tendencies."[37] Likewise, Jones describes how these twentieth-century beliefs grew from nineteenth-century physicians who were prejudiced.[38] Consequently, Jones says, "In this atmosphere it was not surprising that physicians depicted syphilis as the quintessential black disease."[39] According to this view, African American behavior encouraged the spread of syphilis through the black community and presented a threat to white health.

The long-standing influence of racial medicine on the testing and treatment of syphilis in rural blacks clearly influenced the making of this motion picture. The filmmakers were dependent on the medical experts for material and were part of the culture that supported the racist perspectives as a mainstream understanding of the world. Hall is concerned with the "subtle ways in which ideologies of racism" are maintained in society and argues that for "very complex reasons, a sort of racist 'common sense'" develops. The media, he argues, takes this racist common sense as a "baseline without questioning it."[40] Thus, media productions such as *Three Counties* rely on a racist foundation and communicate a racist perspective. Hall argues that the media produce "representations of the social world, images, descriptions, explanations and frames for understanding how the world is and why it works as it is said and shown to work."[41] *Three Counties against Syphilis* presents a racist view of the world predicated on the beliefs of racial medicine that see blacks as prone to diseases, such as syphilis, and therefore a problem population that threatens white society's health.

Although *Three Counties* does not contain the overt racist depictions that are common in *Helping Negroes*, it presents an implicit argument steeped in racist beliefs at the heart of racial medicine, beliefs that were commonly accepted as medical/natural truths. Hall explains that inferential racism is the "apparently naturalized representations of events and situations relating to race, whether 'factual' or 'fictional,' which have racist premises and propositions inscribed in them as a set of unquestioned assumptions. These enable racist statements to be formulated without ever bringing into awareness the racist predicates on which the statements are grounded." Even if the representations are unconscious, they are still racist because of the "assumption that blacks are the source of the problem."[42] In this sense, *Three Counties* is inferentially racist because it is formulated on the racist premises of racial medicine.

It is important to point out that there is no evidence that the filmmakers intentionally presented racist beliefs in this motion picture. As Hall explains, "An ideological discourse does not depend on the conscious intentions of those who formulate statements within it."[43] The filmmakers operated within a racist framework of understanding that inherently believed blacks were naturally a problem population, one prone to immorality and disease among other problems. Thus, the filmmakers did not question racial medicine's suppositions and presented that view.

Of course, many argued at the time that rural blacks had a high rate of syphilis, but racial medicine saw the black community as a problem population because African Americans were inherently inferior. Specifically, racial medicine argued that blacks were morally and physically inferior to whites and that is why they suffered higher rates of syphilis. However, racial medicine ignored the social aspects of black life that were imposed on the black community. *Three Counties* does not locate the problems of the black community within the history of slavery, segregation, and economic despair; instead it divorces blacks from their historical situation and presents their problems as a natural result of their lifestyles.

In *Three Counties against Syphilis* blacks are diseased simply because they are racially inferior, the foundational belief of racial medicine.

Treatment versus Nontreatment

Earlier in this chapter I argued that to fully understand the racist foundations of *Three Counties against Syphilis,* its relationship to the infamous Tuskegee Syphilis Study must be understood. However, the two studies differed in an important way. The *Three Counties* study was a testing and treatment program. Although the film does not go into the ultimate success of the program in treating syphilis, it does present filmic evidence that the PHS attempted to maintain and monitor the patients through the long weeks and months that the treatment required. On the other hand, Fred Gray reports that the patients in the Tuskegee study were not given a treatment schedule and were left to suffer the ravages of syphilis by the PHS doctors who routinely examined them.[44] In fact, in 1939, when a mobile health clinic like the one in *Three Counties* became available in Macon County, Alabama, the nurse in charge of the Tuskegee study was "attached to the unit" to ensure that the men in the Tuskegee Study would not receive treatment."[45] It is clear that, unlike the Tuskegee study, the *Three Counties* program was designed to include treatment to syphilitic residents of Camden, Glynn, and Brunswick counties, in Georgia.

Success of *Three Counties against Syphilis*

Three Counties ends with a final title that states that it was "produced by the United States Department of Agriculture" and credits three USDA Motion Picture Service employees with making the film: Eugene Tucker as the camera person, Raymond Evans (who later became the head of the Motion Picture Service at the USDA) as the editor, and Reuben Ford as the sound person. The USDA archives do not contain evidence that the film received any negative comments or bad publicity. Interestingly, the film stayed in circulation until at least 1951, which indicates it was being utilized.[46] By 1951 the motion picture was thirteen years old, and penicillin was widely available for the treatment for syphilis.[47] It would seem that the film would have outlived its usefulness long before the early 1950s, but the fact that it persisted in the catalog suggests that the film remained a viable resource for the USDA.

Chapter 5

Henry Browne, Farmer

The 1942 USDA film *Henry Browne, Farmer* lasts approximately ten minutes and focuses on Henry Browne and his family, African American farmers living in rural east-central Alabama during World War II.[1] The bulk of the motion picture reveals the Browne farm as a model for all small farmers. The entire Browne family, with the exception of the oldest son, plays an active part in the farming. The younger son milks the cows, raises a calf, and helps with the mules. The daughter raises chickens that produce eggs and meat, and Mother Browne tends a large vegetable garden and puts up canned produce for the winter. Henry Browne is a successful and thoughtful farmer despite the fact that he has a small farm, only forty acres, and that the soil is "not particularly good land."

Thus, *Henry Browne* differs from the earlier USDA motion pictures in its representation of African American farmers. The earlier films either focus primarily on how black farm families need help becoming better farmers and homemakers or present African Americans as a problem race. The present film goes beyond the farm to show the Browne family visiting their oldest son, who is serving in the US military; their son is in training with the Tuskegee Airmen 99th Air Squadron, an all-black military unit. During this segment, the family watches the young cadet as he flies his plane over the base. The film ends with the narrator expressing pride in both the young cadet and his father for their important service to America. The portrayal of the Browne family as good farmers combined with the patriotic service of the oldest son presents a much more positive view of rural African Americans than the earlier USDA motion pictures.

However, these positive aspects of *Henry Browne* ultimately are undermined, because the Browne family's successes take place within the segregated world created by Jim Crow laws and customs that denied African Americans their civil rights.[2] The movie presents the Browne family as good people, good farmers, and good Americans and shows how their hard work made them successful. But the Browne family's lives and efforts are confined to an all-black world. The Brownes exist in a segregated world where blacks and whites do not interact, a black world where black farmers work their land in isolation and black soldiers serve in segregated military units. Thus, the film communicates support for institutionalized segregation and Jim Crow racism as a system that allowed blacks to be successful, as long as they remained in "their place," separate from white society.

The Patriotic Context

Through the visuals and narration, the motion picture establishes that its story is associated with the US fight for democracy in World War II. This is an important point because it sets the tone for the remainder of the film and lays the foundation for its ending. After the title sequence the narrative unfolds in the framework of the national war effort. As the titles fade, the screen is filled with dozens of serious- and determined-looking GIs in full combat uniform marching in close ranks. Their heavy boots loudly stamp out the cadence of their march as they move, like an unstoppable force, across the screen. The narrator, noted African American actor Canada Lee, says, "These are Americans on their way to the battlefronts of democracy," suggesting that these soldiers are on their way to battlefields in Europe or the Pacific.[3] Rather than talk about the battlefields as geographic locations, however, Lee's narration focuses on the importance of the soldiers to the preservation of democracy.

Scenes of the marching soldiers give way to a row of military tanks driving over a dusty road toward the camera. Lee, continuing with the voice-over narration, describes the scene with "fighting men dedicating their lives to securing a better world for all; soldiers of America." The view of the tanks represents US military prowess and further suggests the battlefield and dangers that await the soldiers. Lee emphasizes the phrase "soldiers of America" to extol their patriotic service. The reference to the soldiers dedicating their lives to the war effort underscores their peril and emphasizes the debt the audience owes these men.

The scene changes to a factory welder hard at work, with sparks flying from his welder. Lee says, "But we have another army, the men in plants and factories producing weapons for our fighting forces in an every growing flood, our soldiers of production." Lee delivers the phrase "soldiers of production" in the same serious tone as he delivered "soldiers of America" in the preceding scene. The juxtaposition of the production scene to the battlefield scene links the workers' labors to the war effort. Similarly, the phrase "soldiers of America" is connected to "soldiers of production," linking the factory workers to the soldiers and their mission.

The factory scene dissolves to a long shot of a farm with a tractor driving toward the camera and then changes to a view of a man tending a field with a mule-drawn plow. Lee explains, "And behind all these stands the broad, rich acres of America and the men who grow our food, produce vital raw materials, and give their sons that this land may remain free." The scene fades to black as Lee says, "This is the story of one such American farmer." Thus, the patriotic theme begins its focus on the farm and the efforts of one farmer. Although the farmers are not presented as the soldiers of farming as might be expected from the two previous vignettes, it is noted that farmers give their "sons that this land may remain free." Hence, patriotic service is acknowledged as well as the significance of the farmers' commitment to American democracy.

The Henry Browne farming story is framed in this patriotic context and explicitly tied to American soldiers sacrificing their lives on World War II battlefields. The opening sequence is designed to evoke patriotic feelings and, in so doing, gives the story the weight and importance of the US commitment to protecting democracy and the gravitas of the national war effort. The patriotic context is a powerful rhetorical strategy that establishes the significance of the motion picture to the audience.

The Agricultural Adjustment Administration and Good Farmer Browne

Henry Browne differs from the other USDA farming films analyzed in this book because it does not explicitly mention the USDA extension work with African Americans and it presents Henry Browne as an accomplished farmer, not a black farmer in need of help from the USDA. The film was produced by the USDA Motion Picture Service for the Agricultural Adjustment Administration (AAA), which was established as part of the New Deal. In his book *A Revolution Down on the Farm*, Paul K. Conkin argues that the Agricultural Adjustment Act "proved to be one of the most controversial New Deal bills. . . . The law . . . was among the most complex ever enacted . . . and among the most influential."[4] Conrad explains that its purpose was to increase farm prices, provide some relief to farming indebtedness, and offer other related help to farmers in the economic crisis of the Depression.[5] Agricultural historian R. Douglas Hurt writes, "In addition, direct federal payments to farmers for reducing their production would put needed money into their households and help prevent bankruptcy, increase purchasing power, and keep farmers on the land."[6] *Henry Browne* does not make reference to the AAA in its story and does not comment on the act or its success; however, the significance of mentioning the AAA in the title sequence should not be overlooked because the AAA was important to farming during the Great Depression.

Henry Browne is a farm owner, not a tenant farmer, as was the case with the Collins family in *Helping Negroes* and many of the African American farmers in *Negro Farmer*. This is an important distinction, as historian Van Perkins explains that the AAA's "most outstanding failure . . . was clearly the treatment of Southern tenant farmers and sharecroppers."[7] Indeed Perkins argues that "the tenant problem . . . would never be satisfactorily resolved during the New Deal."[8] Conrad agrees and asserts that "little concern for tenants was shown by the leaders of the Agricultural Adjustment Administration."[9] Likewise, agricultural historian Jerold Auerbach argues that "from the sharecroppers' perspective . . . the [AAA's] most deplorable feature was the disproportionate allocation of benefit payments to landlords. . . . However justifiable New Deal policy appeared to officials in Washington, it enraged the sharecroppers, who hoped for so much and, as always, received so little."[10] Therefore, it is not surprising that the AAA would present the Browne family as farm owners rather than tenant farmers.

The film presents Henry Browne as a good farm owner whose farm is a model for small farmers across the United States. He does not require help from white or black county agents. When a single reference is made to a "government man" who says the United States needs more fats and oils to replace those being used in supporting the war, Browne responds appropriately. As viewers watch Browne, wearing a broad-brim straw hat, hitching reins to his mules, Lee says, "Father's got plenty to do today. It's not that every farmer doesn't always have plenty of work. But this year is different. Different because this year it isn't just cotton and corn." Henry Browne drives his mules from the barnyard toward the fields as Lee continues: "The government man says that we need more fats and oils to make up for what the Japs got. So Farmer Browne thought if peanuts were really needed, he'd try fifteen acres of them. Farmer [Browne] figured, too, it'd be good for a man to try his hand at a new kind of crop. Yes, a good farmer with respect for his land and the job it has to do." Thus, the film presents Browne as a good, responsible, intelligent, patriotic farmer.

Moreover, this scene highlights Farmer Browne as an agent of change on his farm. Unlike the African American farmers in *Helping Negroes*, Henry Browne decides to grow peanuts for the war effort even though it is an unfamiliar crop. In *Helping Negroes*, the power to enact meaningful change on black farms is controlled by whites. In contrast, here it is Browne himself who creates change on his farm. Early in the motion picture, it is established that Browne is a good farmer who understands his farm and his patriotic duty as a farmer and makes a conscious effort to support the war.

Henry Browne differs from the two earlier USDA films analyzed in this book in that the USDA produced the film for the AAA and it presents an African American as a first-rate farmer who understands his farm and his patriotic duty and has the power and will to enact change on his farm.

Content and Stylistic Elements of *Henry Browne, Farmer*

Henry Browne also deviates from the earlier films discussed in terms of its narrative style, narration, and use of diegetic and nondiegetic sound beyond narration. Perhaps the most striking difference is its narrative style. Presented in a story form, the movie presents a day in the life of the Browne family. This storytelling style engages the audience and invites them to be curious about this farm family and their trip to visit their son at a nearby military base. The presentation of the Browne family's story replaces the more objective detached style of the earlier motion pictures with a narrative, inviting the audience to identify with the Browne family as a good American farm family.

Narrator

Henry Browne differs from the earlier films in using the famous African American theater actor Canada Lee, who had a warm, inviting vocal style, to present the voice-over narration. Lee was well known for his stage roles and was a champion of civil rights until his untimely death in 1952. His performance of the voice-over script gives the narration a professional tone and actor-quality presentation that is lacking in the narration in *Negro Farmer* and *Three Counties.* The use of an accomplished actor to present the narration raises the production quality of *Henry Browne* above the other films addressed in this book. Moreover, using a well-known African American actor instead of a white narrator is in keeping with the motion picture's use of only African American cast members to tell Henry Browne's story; thus, all of the principle actors are African Americans.[11]

Diegetic and Nondiegetic Sound

Henry Browne also utilizes a more sophisticated sound track than the other USDA motion pictures addressed in this book. As discussed in earlier chapters, *Negro Farmer* and *Three Counties* use background music in their production, most notably traditional African American gospel hymns in the closing sequences. *Henry Browne* not only uses background music throughout its production but also adds sound effects at appropriate points in the film. The background music heard throughout the film is incidental instrumental music much like that used in mainstream Hollywood productions of the time. The score is upbeat and suggests a country or rural theme. Perhaps more significantly traditional African American gospel music was not featured, as it was in both *Negro Farmer* and *Three Counties.*

Neither *Negro Farmer* nor *Three Counties* use diegetic sound in their sound tracks. *Henry Browne,* however, uses postproduction sound to emphasize important narrative elements in the film. In the opening patriotic sequence, the sounds of the soldiers' combat boots loudly stamp out the cadence of their march to emphasize their travel to the battlefields of democracy. At the end of the film, the sound of the airplanes' engines is re-created on the sound track as the planes taxi out, take off, and fly over the base. The heaving droning of the engines dominates the sound track and adds an impressive stylistic element to the film. The engines' roar as the planes pass over is reminiscent of similar scenes in Hollywood war films.

By employing a storylike narrative style, a professional African American actor as the narrator, and creative use of sound elements, *Henry Browne* stylistically differentiates itself from the earlier USDA motion pictures analyzed in this book. The higher production values and greater use of the stylistic conventions of Hollywood cinema make *Henry Browne* a more polished and entertaining motion picture. The more entertaining narrative style invites the audience to positively identify with the Browne family story.

Constructing the Reality of the Browne Farm

As with the other films addressed in this book, the filmmakers constructed a reality that informs and persuades the audience to understand and accept the content as accurate and believable. *Henry Browne* accomplishes these goals by showing the Brownes as they go about their lives on the farm; segments include the introduction of the members of the Browne family and their farm, the family eating a nutritious breakfast, the merits of the family garden, the importance of raising livestock, the preventive maintenance and reusing of materials, and the careful use of the land. Each of these segments demonstrates that the Brownes are model farmers who utilize the best judgment and practices in managing their farm and home.

Introducing the Browne Family and Their Farm

Following the opening patriotic sequence, the film introduces the audience to the Brownes and their farm in order to show that the Browne family is a typical American farm family. The audience first glimpses the farmstead's wooden barns. The camera then pans across the yard to reveal a modest wooden farmhouse. The view is of a common farmstead that could exist almost anywhere in the United States.

From this vantage point, viewers see an African American man standing on the porch of the house. The film cuts to a medium close-up of the man, wearing a light-colored shirt and overalls, as he washes his face in a water bowl on the porch. Lee introduces the man: "This is farmer Henry Browne, age thirty-eight, father of three children, farming forty acres. This year Henry Browne is growing more [crops] than ever before." Farmer Browne finishes washing his face and pours the water out of the bowl. Lee continues: "And to get the work done these fine summer days, he and his family have to be up with the sun." This brief scene efficiently accomplishes a great deal in a short time span. Whereas the title of the film avoids explicit reference to African American farmers, unlike *Helping Negroes* and *Negro Farmer*, the visual images in the film's introduction establish that Henry Browne is an African American. Throughout the film no reference is ever made to Browne or his family members as being African Americans. They are always simply referred to as farmers, not *Negro* farmers (as in the earlier USDA movies).

The Nutritious Breakfast

After introducing Henry Browne, the scene changes to show the Browne family having breakfast. This scene communicates to the audience that the Browne family enjoys a nutritious diet that is free of the problems experienced by black farm families in the earlier USDA film *Negro Farmer.* The scene opens on a black woman wearing a dress and apron and working in the kitchen. Henry Browne is seated at the table eating breakfast. A boy, Young Henry, enters and sits at the table next to his father. A pitcher of milk is seen on the table. Lee proclaims, "Milk for breakfast and eggs too." Panning right, the camera reveals a young black girl also eating at the table. Lee focuses on her importance to the breakfast scene: "Sister is raising twenty hens, most all good layers. 'Those that aren't good layers will be good eaters,' she says." The breakfast segment establishes that the Browne family lives in a modest farmhouse and enjoys a healthy diet thanks to their livestock efforts.[12]

The Browne family breakfast is typical of hundreds of similar breakfast scenes in Hollywood films and communicates normality and ordinariness in its familiarity. This sense of ordinariness is important because the previous motion pictures were focused on the problems related to black farm families, the emphasis being on their differences from white people. By replacing the nutrition problems attributed to the pellagrous diet of African American farmers in *Negro Farmer*, with this scene of a healthy breakfast, the film emphasizes that the Browne family does not suffer from the problems the USDA touted as black farm family problems in its previous films. The contents of the breakfast are significant. Lee comments that there are milk and eggs for the family breakfast, a vastly more nutritious meal than *Negro Farmer* suggests black farm families consume. Lee comments that the Brownes' breakfast has improved in the last year: "Last year there was just the same old cornbread and fatback for breakfast. Fills you up all right, but it didn't build strong, healthy bodies. Young Henry grows like Johnson grass [a thin, fast-growing weed that grows several feet in height], but even so he's already begun to fill out." The motion picture does not explicitly address *why* the Brownes' breakfast improved in the last year, but it implies that it is because Sister is raising hens that lay eggs and because the family now has a milk cow. The movie does not explain why the family has a cow and hens this year (and not in the past), but it also does not suggest that the USDA improved the family's circumstances. The audience is thus left to assume that the family itself is responsible for these improvements.

The family breakfast scene creates a cinematic reality of health, well-being, and success for the Brownes; it suggests they are fit and enjoy a nutritious diet that includes healthy options such as eggs, meat, and milk. Moreover, unlike earlier USDA representations of black farm diets, this breakfast capitalizes on its ordinariness to emphasize that life is good for the Browne family.

The Browne Family Garden

Following the breakfast segment, the film takes the viewer to the Browne family garden, where Mother Browne works diligently. The voice-over explains, "Before going out to help with the cotton, Mother tends her garden a bit. There's not much here that will be sold, but there's a lot here that will be eaten this summer and some that will be put away for next winter." The scene continues with close-ups of several vegetables being grown in the garden, including tomatoes, watermelons, and peas. Lee says, "No, sir, the Brownes won't be depending on a grocery store. They'll leave canned things for people in the cities, who just can't grow their own vegetables." Thus, through Mother Browne's efforts, the garden provides nutritious fare for the family, and the Brownes help the war effort by not consuming canned foods that are needed to feed city dwellers. In this way the garden is both an important addition to the Browne family provisions and a source of pride for the family as it helps them to further contribute to supporting the war.

A comparison to *Negro Farmer* is appropriate for the garden sequence also. In *Negro Farmer* the need for a family garden is stressed as being fundamental for black farmers, and its benefits are lauded. In *Henry Browne*, on the other hand, the Brownes' lush, healthy garden is used to demonstrate another facet of being a successful farmer. By maintaining an abundant garden, the Brownes have plentiful food for themselves and little need to buy canned food, which is needed for consumption by families that are unable to grow their own food. Thus, the garden contributes to the cinematic reality of the Brownes as an accomplished farm family that contributes patriotically to supporting the war by growing their own food.

Raising Livestock

Raising livestock is touted as an important fundamental to the black farm in *Negro Farmer,* and in *Henry Browne* we learn that the Browne family is accomplished in raising livestock. For example, the audience is introduced to Henry Browne's son, Young Henry, through the boy's work with the family's cows. The scene opens with a young black male, barefoot and dressed in a light-colored shirt and overalls, milking a cow. Lee introduces him in voice-over narration: "This is Young Henry doing the milking. This year, for the first time, they have a cow, and Young Henry is raising the calf as his own." On cue, a calf walks into the scene. Lee proudly states that the calf is "going to be the best calf in the county." Thus, the motion picture establishes Young Henry as a willing, active, and successful participant in family farming and points out that the Browne family is raising livestock. Young Henry's active participation is a vastly different presentation from that of the passive children represented in the poor housing sequences of *Negro Farmer.* Young Henry is both active and productive in his work and in his contributions to the family farm and livelihood.

Sister Browne is equally dynamic and industrious. During the breakfast sequence, Lee explains that the breakfast eggs are available to the family because "Sister is raising twenty hens, most all good layers. 'Those that aren't good layers will be good eaters,' she says." This brief remark suggests that Sister helps with the family nutritional needs by providing fresh eggs and meat. Later, the film revisits Sister's work as the audience views Sister feeding her hens in the henhouse and yard. Lee comments, "Sister is a pretty good farmer too. She certainly works hard tending her chickens; must be worth the trouble, though, because she has a mighty healthy flock," as the viewer is treated to a scene of many healthy chickens. Sister's work with the hens and Young Henry's work with the cows is reminiscent of scenes in *Negro Farmer* in which young blacks are shown as productive and helpful farm family members when properly influenced by the USDA extension workers through approved programs of the 4-H club. The scenes of Young Henry working with the family cows and Sister working with the hens communicates the filmic reality that the Browne family farm is an ideal farm in which the children contribute productively to the health and well-being of the family and the success of the farm.

Avoiding Waste and Employing Preventive Maintenance

The reality that *Henry Browne* works to create for the audience is that Henry Browne is an excellent farmer and that his family contributes positively to the farm. Farmer Browne's virtues are demonstrated in a sequence that shows how he avoids waste and practices careful preventive maintenance to protect and extend the life of his farm equipment.

This sequence begins with an establishing shot of the Browne family barn and moves to focus the audience's attention on a close-up of neatly stacked empty burlap fertilizer bags. Canada Lee comments, "Doesn't seem like much, but these are the sort of things that make a [good] farmer. Like putting a new board on the barn when it's needed and saving those burlap fertilizer bags last fall. Now we need them bad, and we're glad they weren't thrown away." Lee's remarks are not limited to the Brownes alone. Lee encourages the viewers to identify with Farmer Browne's good judgment when he says, "we need [the burlap bags] bad and we're glad they weren't thrown away." Lee's uses "we" to mean all Americans because everyone can help fight material shortages resulting from the war. Thus, saving burlap bags is not only a wise choice for Farmer Browne but also a patriotic act because it helps with material shortages caused by the war.

Immediately the audience is treated to a series of shots of farming equipment (e.g., plows) as Lee continues: "And because his farm machinery was greased and put under cover when not in use, it will probably last for the duration [of the war], and that's important because as long as our factories are turning out tanks, we just have to make everything last." Lee's narration makes the patriotic connection explicit in this scene. Farmer Browne's attention to proper preventive maintenance of his metal farm apparatus means the equipment will last longer; his efforts directly aid the US war effort because the factories needed to focus on constructing tanks.

Farmer Browne's decisions to save and reuse everyday materials, such as burlap bags, and to maintain his farm equipment contribute to the portrait of Farmer Browne as a good farmer and a patriot whose attention to even the small aspects of his farming operation maximize his contributions to the war effort.

The Land

The motion picture continues to present the Browne family's patriotism with a scene of Henry Browne and his mules arriving at a field where his plow is waiting. Browne bends over to check the soil and then looks up to scan the sky. Lee explains in voice-over narration, "Last fall he [Henry Browne] put lime and phosphate on twenty acres of the poorest part of his farm, and now both he and his country will benefit from the better crop he'll harvest. If the weather holds, Henry figures on making a really fine crop of peanuts," as the scene fades to black. The reference to the peanut crop reminds the audience that Browne planted peanuts this year to help with the shortage of oils needed for the war, reinforcing the view of Henry Browne as a patriot because he is using part of his land to grow a crop that is needed for the war effort.

The land sequence is expanded with a scene of Mother Browne walking from the barnyard toward the field carrying a long-handled hoe. Sister Browne runs into the scene, grabs a hoe, and follows her mother to the field. They walk to the field together and begin hoeing between the rows of crops as Lee says, "Forty acres of land—not particularly good land either." As the audience sees a shot of the barn, Lee continues his narration: "A barn that isn't big or even new, and no tractor." The reference to the lack of a tractor bridges the transition to a scene of Henry Browne plowing with his mules. Lee acknowledges the mule team: "Just a team of mules, an understanding of the earth and things that grow, and long hours of hard work for everyone." In this way the motion picture concomitantly emphasizes the Browne family's modest resources, hard work, and success. The audience is invited to identify with the Brownes not as prosperous landowners with modern farming equipment but as a farm family of common means who are making the most of what they have to be efficacious farmers and patriots.

Lee continues to extol Farmer Browne's virtues as the viewer enjoys a scene of properly plowed acreage with neat rows running along the contours of the land. As the scene changes to close-ups of the crops in the field, Lee says, "Even the land . . . can be wasted. If Farmer Browne hadn't planted those peanuts in rows following the contours of the slopes, rain could wash away both his crop and his land. And right now this land of Farmer Browne and all the other farmers is mighty important to our country, because we need more peanuts for oil, more long-staple cotton, more food. So it's up to everyone to make the most of every acre of land and every bushel of seed that goes in it." Hence, Farmer Browne's use of proper plowing techniques to ensure that the crops are planted along the contours of the land to avoid soil erosion is the capstone of the motion picture's construction of the Browne family and their farm as being patriotically successful. Moreover, this sequence ends with Lee's appeal reaching out to the farming audience to remind them that, like Henry Browne and his family, they too have an obligation to make the most of their resources in order to support the United States during the war.

The Tuskegee Airmen

After demonstrating the Browne family's successful and patriotic farm practices, the focus shifts to Henry Browne's oldest child, who joined the Tuskegee Airmen in order to fight in World War II. The importance of this sequence is that it links the farm to the war both visually and narratively. In addition, these scenes reveal to the audience that the Browne farm is located in Macon County, Alabama, home of Tuskegee Institute and the all-black 99th Pursuit Squadron.

The film reveals the Browne farm location in a subtle manner, unlike the overt measures used to establish locale in the three previous USDA films. The narrator temporarily abandons the omniscient perspective and adopts the viewers' perspective so that for several minutes he appears to be a viewer of the action. This part of the film opens with a scene of the Browne family climbing into their mule-drawn wagon. Lee says, "Saturday afternoon and, of course, this is a day that farmers go to town just to sell some eggs that Mother and Father market." But his next statement arouses curiosity: "Even so, it's early. There must be something special today." Lee's questioning tone signals to the audience that indeed something special should be expected. Narratively, the special event is a visit with the older Browne son. Rhetorically, the special event is the further linking of the farm to the war.

To this point the locale could be almost anywhere in the southern United States, but after the Brownes travel through the countryside and enter a town, the location becomes clearer. The family passes by the W. J. Brantley grocery store, then the Macon movie theater, and later a large church. The narrator wonders aloud why the family does not stop in town: "Aren't they gonna stop? They're going right through town." Although the film never explicitly discloses the location, film scholar Peter Noble later revealed that the movie was filmed in Macon County, Alabama.[13]

The scenes that follow take place at the 99th Pursuit Squadron military base and were filmed near Tuskegee, Alabama. The 99th Pursuit Squadron, better known as the Tuskegee Airmen, trained first at Moton Field in Macon County, Alabama, and later at the Tuskegee Army Airfield.[14] Given the year the film was shot, these scenes were most likely filmed at the airfield.

Narrator Lee maintains the spectator's point of view as the Browne family leaves the city, and he heightens the curiosity about the family's undisclosed destination by asking, "What's all this about? What's got Sister so excited?" Then in cinematic answer to Lee's question, the mystery is solved as the camera reveals a soldier at a military base guard shack allowing a military jeep to leave the base through the gate. Lee chimes in: "Well, no wonder they didn't stop in town. No wonder those young ones are excited. It's a visit to big brother, who is a cadet training with the Ninety-ninth Pursuit Squadron." This sequence establishes the geographic locale of the Browne farm as Macon County, Alabama, and the fact that Cadet Browne is training with the Tuskegee Airmen. The geographic proximity of the farm and the military base bolsters the rhetorical linking of the farm to the war because the farming and military training take place in the same county. Additionally, the county is home to the historic Tuskegee Institute, famous for its agricultural education and outreach.

Having established this geographic connection between the farm and the military base, the next sequence emphasizes the Browne family's commitment to the war. In the sequence Cadet Browne takes part in flight training with other members of the all-black 99th Pursuit Squadron.

As the wagon approaches the military guard post, a young black soldier in a flight suit walks up to the wagon and climbs aboard. This is obviously the oldest Browne child. The next camera shot is a medium close-up of his smiling face framed by those of his younger sister and his mother, who are both very pleased to see him. The soldier hugs his younger sister and shakes hands with his father. Lee provides cinematic foreshadowing as he says, "This is something special, all right. Maybe there will even be a chance to watch him fly." This foreshadowing is important not only to the motion picture narrative but also to audience members who are interested in seeing military planes.

As the Browne family is reunited on the wagon, the military guard hands the cadet a pass, and the Brownes drive their wagon on to the base and park it next to a military truck. The symbolism of having the rural farm wagon parked next to a military truck furthers the visual connection between the farm and the war.

Lee falls silent and allows the audience to witness the next events without voice-over narration. As the background music plays, audience members are allowed to interpret the visuals for themselves. The film uses this sequence to emphasize the location as a military base and to demonstrate that Cadet Browne is an active participant in this scene while also showing the audience many African American military personnel in the process. The five Brownes walk toward the airfield across the base as military trucks and jeeps speed past. Aircraft hangers can be seen in the background as the Brownes make their way to their destination. Soon the family reaches the airfield and the planes. A series of shots show the cadet pointing out and explaining things about the planes and the airfield to the family. The family members are shown asking the cadet questions, and he is seen answering, but no voices are heard on the sound track. As the family talks, another African American airman walks up to the group and hands Cadet Browne some flight equipment and helps him into his parachute harness. The introduction of the second pilot signals the end of the family reunion, as Cadet Browne will once again be visually separated from the Browne family. Cadet Browne, a beloved family member, leaves his family to accompany his fellow airman to the military planes. Thus, the Brownes symbolically release their son to the war effort.

The two airmen turn from the family and walk toward the airplanes. Both soldiers stop in front of an officer and salute before continuing. Shots of the Browne family's facial expressions show their pride as Cadet Browne leaves them to rejoin the military activities. The camera focuses on Cadet Browne, showing him climbing into the cockpit of his plane and readying it for takeoff. The nondiegetic music ends, and viewers hear the sound of the planes' engines as they roar to life and their propellers whirl. The camera switches to a long shot of the runway as three airplanes take off together. The camera switches to a view of the Browne family in the foreground watching the planes take off and fly overhead. The camera remains on the Brownes as they watch the airplanes fly out of sight. The entire airfield sequence is presented without voice-over narration. This scene of the Brownes sending their son overseas for the benefit of American democracy joins the farm to the battlefield and the farmers to the fighting.[15]

The Patriotic Appeal

Having established that Cadet Browne is serving his country, the motion picture comes full circle with a patriotic appeal that all Americans should be grateful to the Brownes for their patriotic duty on both the farm and in the military. Thus, patriotism is used to reinforce the message that the Brownes are good farmers and good Americans. *Henry Browne* began with a military context explaining the importance of America's soldiers, factory workers, and farmers to the war effort. During the farming vignette the film notes that the farmers give their "sons that this land may remain free." Cadet Browne is the cinematic exemplar for all farm families who give their sons to the war effort.

Canada Lee's narration begins anew as the camera focuses first on the faces of Henry Browne and his wife as they look skyward at the disappearing fighter planes: "Farmer Browne and his wife are mighty proud of their son, and so are we. But we are also proud of Farmer Browne. Proud of him because he is doing his job, an important job, raising more livestock, saving his land, and growing more things we need. And proud of him for just being an American we can count on at a time when every American has an important job to do." With this, the background music swells up, and the scene fades to the end title with Lee's voice-over narration nicely summarizing the rhetorical point of the film for the audience. As Lee explains, Farmer Browne is a good American on whom all Americans may depend because he does what is needed to further the war effort on his farm and because his son is taking the fight for democracy to the enemy. Thus, like filmic bookends, the military segments frame the farming sequences and make the connections explicit for the audience. Farmer Browne is a good American who is doing what is right for America so that all Americans can benefit.

The Use of Stereotypes in the Film

Henry Browne breaks with the use of stereotypes seen in the earlier USDA movies analyzed in this book. The film avoids the blatant stereotypes seen in *Helping Negroes* that define African Americans as different and a problem race. The most interesting aspect of the film is its attempt to avoid such racist depictions. Historian Thomas Cripps explains in his book *Slow Fade to Black* that *Henry Browne* was one of the films that "contributed to the destruction of the old monopoly of Southern racist attitudes on the screen."[16] In *The Negro in Films*, Peter Noble explains that *Henry Browne* "depicts a small American farmer, a Negro, at work in the fields and at home, emphasizing his contribution to the war effort. It shows the farmer's family working in the fields with him, and finally we see them all going into the nearest town to visit the oldest son, a soldier-mechanic in the American Air Force. As Iris Barry notes in a letter to the author [Noble], 'This film is effective because these people 'simply happen' to be Negroes, and no particular point is made of this.'"[17]

In many ways *Henry Browne* purposefully works to contradict stereotypes used in the earlier USDA motion pictures. It does not contain scenes of African Americans dancing to fiddle music or eating watermelon. The Browne family members are African American, but the movie does not refer to them as Negroes (as in the earlier films). The 99th Pursuit Squadron is an all-black military unit but is not referred to as such. In this way the film strives to avoid direct references to Negroes (common in the USDA motion pictures) and establishes that the film is about an American family that just happens to be African American.

Moreover, the film does not present blacks as a problem population. African Americans are not represented as needing government help, as diseased, or as a threat to white society. The Browne farm is represented as successful and self-sufficient, unlike earlier depictions of black farms in USDA films. The farming and homemaking aspects of the Browne family farm are exemplars (from the USDA perspective) for other small farm families with a nutritious garden and healthy livestock. The family members are active, and Farmer Browne is an agent of positive change and shows excellent judgment in all of his decisions. From these perspectives, *Henry Browne* presents African Americans in a positive manner. Indeed, the film elicited positive responses; Cripps writes that the motion picture "seemed so powerful that Lawrence Reddick, curator of Harlem's Schomburg Collection, screened it to cool down the street crowds after the Harlem race riots in 1943."[18]

Henry Browne was much different from *Negro Farmer*, released just a few years earlier. The film's intentional avoidance of many stereotypes and its attempt to show that the Browne family members were good Americans and good farmers diverged dramatically from earlier portrayals of the black community as a problem and in need of white care and control. It is obvious that the USDA filmmakers took care not to continue the use of themes suggesting that blacks were unintelligent, lazy, immoral, prone to disease, or a problem race. Despite these efforts, *Henry Browne* left the racist fundamentals of segregation unchallenged and unquestioned.

By doggedly clinging to the racist ideology of the separation of blacks and whites, *Henry Browne* fails in its effort to depict the Browne family as just plain farmers, not *Negro* farmers. The Brownes live in an all-black world brought on by the pervasive racist practice of segregation. For example, the Tuskegee Airmen was an all-black military unit because of institutionalized segregation.

Jim Crow and *Henry Browne, Farmer*

Ultimately, despite the filmmakers' choice to avoid stereotypes, *Henry Browne* relies on the fundamental aspect of Jim Crowe racism, that blacks must be separated from whites in the United States. The film avoids referring to the idea that blacks are different from whites by simply avoiding interaction between whites and blacks and focusing exclusively on African Americans. Therefore, even though its treatment of African Americans is more positive than the earlier USDA films that dealt with blacks, the film supports segregation.

Since *Henry Browne* is only about black farmers and black soldiers; it remains fundamentally about race while simultaneously avoiding explicit reference to race. Given the constructed reality that *Henry Browne* creates for the viewer, analysis of the motion picture does not suggest an overall positive reading. Despite its improvement in the representation of blacks from earlier USDA films, *Henry Browne* was not universally embraced by blacks. In "*The Negro Soldier* (1944): Film Propaganda in Black and White," Thomas Cripps and David Culbert argue that *Henry Browne* "failed to convince anyone that racial tolerance was desirable. Browne was the perfect obedient Negro: possessor of 40 acres, some chickens, a son in the black 99th Pursuit Squadron, and a willingness to grow peanuts because his country needed their oil." Moreover, Cripps and Culbert explain that the African American journalist "who originally suggested the idea [for the film] termed the finished product 'an insipid little story.'"[19] It is not surprising that some viewers found *Henry Browne* dull and uninteresting; with the exception of the scene where the airplanes take off, the film's tone, pace, and subject matter are bucolic, and the story is neither gripping nor thrilling.

The central problem with the film is not just that it communicates the acceptance and maintenance of the separation of white and blacks in America, but also that it celebrates that separation. The reality of the motion picture is that Henry Browne and his family are good farmers and good Americans, and it appears that the films considers it good that they are completely separated from their white counterparts. In this way, the movie supports Jim Crow racism by suggesting that it is working in favor of blacks in the United States. The Brownes appear to suffer no problems related to segregation, and, more importantly, segregation appears to work very well, despite the fact that Jim Crow laws were oppressive and denied African Americans their civil rights. Farmer Browne and his family are fine and are doing their part, within the limits of Jim Crow, to contribute to the war effort. Thus, the film argues for racial separation, not racial tolerance or integration, and suggests there is no need for racial tolerance as long as the Brownes and all other blacks remain in their place and separate from whites. Despite the racist treatment of African Americans in *Three Counties*, that film at least implies that blacks and whites interacted. In contrast, *Henry Browne* suggests that the Browne family is able to function very well within the Jim Crow reality of 1940s Alabama with no difficulty resulting from institutionalized racism. What need, the film argues, is there for racial tolerance when racial separation is the key?

The Brownes are represented as happy, healthy, productive Americans who are supporting the war effort with both their farming and their son's military service. They do not have to interact with whites to do so. Their filmic world is populated by blacks; even the military is segregated. Their home is not the ramshackle shanty of *Negro Farmer*, and they do not require treatments for the "bad blood" as *Three Counties* suggests is needed for African Americans in the rural South. There is no need for county extension workers to help them become better farmers and homemakers because they are already model small farmers and homemakers. Hence, the Brownes live in a black America where everything appears to be fine, beyond the external threat of the Axis powers.

The Brownes live in a black America that is separate from white America, and the film suggests that the situation is not just okay; it is good. I argue that the film is deceptive in its support of the doctrine of separate-but-equal Jim Crow racism. The Brownes thrive in a racially separate world where everything is not only going well but also improving. In the last year the Brownes have improved their meals, livestock, and crops. In comparison with the other USDA films about rural black populations, *Henry Browne* presents favorable conditions for African American farmers in Alabama.

The separation of the Browne family, both via Jim Crow laws and the choice to present them outside the context of a farm community, helps the filmmakers avoid situations where they must show blacks as inferior to whites. Concomitantly, it forces the Brownes into a type of cinematic jeopardy where they can be productive and successful—but only in a framework of racial separation. Moreover, the motion picture does not suggest that the Brownes are part of a larger black community. Unlike the earlier films, there are no scenes of rural blacks talking at a local store, participating in communal efforts to can foods, or attending church services. The result is that the Browne family exists outside of a broader community. In *Three Counties* a focus was on the importance of the black community, particularly the black church; the complete omission of such a communal group seems odd in *Henry Browne.* The Brownes are thus not only separated from white society, but they also appear to exist separate from a black community.

In some ways *Henry Browne* is the filmic realization of black extension agent F. L. Blackshear's 1919 admonition that black county agents "avoid mooted religious, political or racial problems" when talking with black farmers to avoid raising the ire of whites.[20] Blackshear suggests that only judicious statements should be made about the race problem and then only with the goal "to make peace between the races."[21] He basically argues that black agents should stick to talking about farming and avoid any topic that might cause whites to be suspicious of the USDA's work with African Americans. *Henry Browne,* made many years after Blackshear's warnings, takes his advice seriously and presents the Browne family as farmers who "simply happen" to be black and who pose no threat to white hegemony because they are happy and successful in their segregated world.[22]

As discussed in chapter 3, *Negro Farmer* ends with the narrator voicing selected parts of Booker T. Washington's "The Case of the Negro," published in 1899.[23] The selections that the USDA filmmakers choose include Washington's statement that "since the bulk of our people [African Americans] already have a foundation in agriculture, [they] are at their best living in the country and engaged in agricultural pursuits."[24] In *Henry Browne* the USDA takes Washington's argument and presents it visually with the Browne family, using the narrative as a way to rhetorically argue that blacks are successful in the country working on the farm. As I explain in my analysis of *Negro Farmer,* the USDA uses these phrases from Washington's article to argue that blacks are most prosperous on southern farms. Although the main thrust of *Henry Browne* may be to demonstrate the Browne family's patriotism, it also is the only USDA film in twenty years to present a black farm family that succeeds without the overt intervention of the USDA. However, that success is shown to be happening only in the segregated world of Jim Crow. What the USDA filmmakers do not quote from Washington's article are his arguments against the disenfranchisement of blacks in America. Washington states, "It is not best for the Negro that he relinquish any of his constitutional rights; it is not best for the Southern white man that he should."[25] Yet the fictional Henry Browne and his family live in the same segregated world as their real counterparts, where Jim Crow laws deny blacks their civil rights through the doctrine of separate but (un) equal.[26]

Author Jerrold Packard explains that "by the early years of the twentieth century the forces responsible for it—white power, white supremacy, and Jim Crow—had returned the large majority of African Americans to a status not unrecognizably different from that they had suffered as slaves, and left the country in a position where racism provided the dominant social undertone of American life."[27] In the final analysis, *Henry Browne* not only represents that reality but also encourages the audience to perceive it as good.

Success of *Henry, Browne, Farmer*

Despite its support of the doctrine of separate but equal, *Henry Browne* enjoyed much success with audiences. The materials in the USDA archives suggest that the film was appreciated by both black and white viewers. For example, a 1943 conference paper written by Chester Lindstrom, associate chief of the USDA Motion Picture Service, explained that *Henry Browne* is a good example of a wartime morale-building film that "shows a representative Negro farm family doing its part in the agricultural war production program, while a son trains with the 99th Pursuit Squadron near Tuskegee, Alabama. Though made primarily for Negro audiences it is also popular with [white] people."[28] Likewise in another letter, the assistant to the USDA secretary explained that *Henry Browne* "is primarily for presentation to audiences of Negro farmers. . . . [But] the picture was so well regarded that one of the distributing companies asked for rights to show it in theatres on a non-profit basis, and has done so. In addition, about 35 prints are being used by State USDA War Boards or otherwise by Department agencies. The Motion Picture Bureau of the Office of War Information has purchased 290 prints and allocated them to various non-theatrical distributing depositories throughout the country."[29] These documents show that *Henry Browne* received wide distribution beyond the regular USDA channels and was shown in commercial theaters.

The USDA also received positive reports from its own distribution of the film. For example, in a 1943 field report on the use of Department of Agriculture motion pictures, agents report that "'Henry Browne, Farmer' has proven popular with Negro audiences in Tennessee. It is impressing the Negro farmer with the fact that however small an operator he may be he can do his bit."[30] Thus, black farmers in Tennessee responded well to the motion picture as a morale builder as Lindstrom had indicated. Likewise a report from Oklahoma reads:

> Another film which has come to our State is the Negro picture, "Henry Browne, Farmer." The first time it was screened was at a Negro agricultural leaders' conference at Langston University. When the lights were turned on immediately after the last scene, there were tears running down the cheeks of many [people in attendance]. The Negro Extension District Agent asked that we lend him the film so that he could show it to his people all over the State. And from the reports which have come in, he has done a good job of showing it. For instance, during the month of December 1942 the picture was shown seven times to 1,312 people.[31]

This report indicates that the black extension agents in Oklahoma regarded the film highly and showed it widely to their constituents. It is clear that the USDA was pleased with the reception of *Henry Browne* by both white and black audiences.

More interesting to this analysis, however, is a positive review of the film sent to the USDA by a leader of the National Association for the Advancement of Colored People (NAACP). In a letter to the USDA, NAACP secretary Walter White praised *Henry Browne*. Secretary White expressed the NAACP's "gratification at the making of this excellent film picturing the participation in the war effort of an admirable Negro farmer and his family. This should be of interest and value both in raising the morale of Negroes themselves and even more in the emphasis to white audiences of the varied ways in which Negroes are helping America during the war crisis. . . . May I express the hope that other films will follow including those which will show Negroes and whites working together as fellow Americans."[32] Secretary White compliments *Henry Browne* for its patriotic message and its ability to educate white audiences on the value of black farmers to the American war effort. However, he also implies that the film is lacking in its failure to show blacks and whites working together. He realizes that the motion picture's major flaw is its reliance on the Jim Crow separation of the races.

Henry Browne did not receive universal acclaim; as mentioned earlier, Cripps and Culbert report that one black reporter complained that the motion picture was "insipid."[33] Nevertheless, the USDA received favorable feedback from a variety of sources, including the NAACP, that indicate the film was successful in its goal as a wartime morale-building film that was accepted positively by both black and white audiences and considered a success by the USDA.

Chapter 6

USDA Motion Picture Representations of African Americans

In this book I analyzed four USDA motion pictures that deal explicitly with African Americans: *Helping Negroes to Become Better Farmers and Homemakers* (1921), *The Negro Farmer* (1938), *Three Counties against Syphilis* (1938), and *Henry Browne, Farmer* (1942). Each analysis included an in-depth investigation of the reality created by the filmmakers, the motion picture's use of stereotypes, the rhetorical message communicated, and a discussion of each film's situation within broader racist ideologies. In this final chapter I build on these analyses to show how the motion pictures communicated the racist ideology that African Americans are inferior and a threat to white society.

Helping Negroes reflected the racist doctrine of separate but equal in rural Alabama in the early 1920s while also positioning black tenant farmers as helplessly unable to improve conditions on their farms. The motion picture rhetorically argued that white authorities such as USDA officials, extension agents, and landowners had the power and knowledge to enact positive change for black tenant farm families. The result shown in the film was that the black tenants' lives were dramatically improved thanks to the help of white authorities. Moreover, the film downplayed Tuskegee Institute's important role in establishing, implementing, and maintaining black extension work in Alabama by attributing the impetus, organization, and success of black extension work in Alabama to white extension agents and officials. In maintaining polarizing differences between blacks and whites, the motion picture supported separate-but-equal laws and customs and communicated that blacks are inferior to and dependent on whites.

In 1938 the USDA released *Negro Farmer,* its second film devoted to explaining how the USDA extension service helped African Americans. *Negro Farmer* presented black farmers as desperately needing help in practically every aspect of their lives from nutrition to sanitation, gardening to raising livestock, and demonstrated how their lives could be improved through extension demonstrations. The motion picture also supported the doctrine of separate but equal while adding a strong argument opposing African American migration from southern farms with selected quotes from Booker T. Washington's 1899 essay, in which he encouraged blacks to remain engaged in agricultural pursuits. Although never explicitly stating that African Americans should remain on southern farms, the position was made clear through a sequence that highlighted Washington's words and bolstered the argument with gravitas by including a famous African American gospel hymn sung by the Tuskegee Institute Choir in the sound track. By arguing that southern blacks should remain on the farm, the USDA's message also supported the racially oppressive system of segregation that forced African Americans into a state of economic peonage after slavery.[1] Thus, *Negro Farmer* bolstered Jim Crow laws and practices while explicitly arguing that blacks should remain on the farm.

Three Counties, also released in 1938, presented the story of a syphilis testing and treatment program for African Americans that took place in three counties in southeastern coastal Georgia at the same time the now infamous Tuskegee Syphilis Study was being conducted in Macon County, Alabama. Although in its early moments the motion picture suggested that the treatment program was for whites and blacks in any community, it quickly abandoned this position and focused solely on the testing and treating of African Americans for syphilis; syphilis was thus portrayed as a black disease that threatened white health. The film communicated support for the program from the black community and the black church. However, like the Tuskegee study, the program supported the ideological position of racial medicine that defined blacks as a problem population, physically and morally inferior to whites, and a threat to white health through social contact. Thus, the motion picture maintained support of segregation while also emphasizing the perceived medical threat to whites by African Americans.

In 1942, during World War II, the USDA released *Henry Browne, Farmer.* This film differed from the other films discussed here in that it avoided representing African Americans as dependent upon whites for help by portraying the Browne family as successful small farmers and their farm as a model of good farming practices. The movie also showed how Farmer Browne's good farming techniques contributed to the war effort in many positive ways. In addition, *Henry Browne* moved beyond the farm to show Henry Browne's oldest son serving in the US military with the

Tuskegee Airmen. The patriotic tone and cinematic reality presented in the motion picture rhetorically argued that Henry Browne was a good American and a successful farmer. However, by completely separating the Browne family's accomplishments from white society and segregating them in an all-black world, the film argued that the racist ideology behind Jim Crow, which denied African Americans their civil rights, not only was acceptable but also worked well for African Americans who were willing to remain on the southern farm and in their segregated place.

Communicating Three Themes

Taken as a whole, these films presented the USDA's cinematic view of black Americans. The importance of these films should not be underestimated; they were seen by millions of people both in the United States and abroad for many decades. Carrying the USDA's official approval, the films intended their audiences to perceive them as accurate and objective representations of reality. The authority of the USDA on matters related to US agriculture was unquestionably strong because it could focus considerable resources on problems and issues facing the American farm. The fact that the USDA invested time and government resources in making these films points to their importance to the USDA and its desire to communicate its message to the American public and foreign audiences.

Moreover, the USDA MPS understood very early in its existence that the major value of its motion pictures lay in their propaganda potential. The USDA saw these films as a way to create positive public relations for their institution and to build public support for their efforts. The films argued the USDA's position on issues important to the USDA and carried its message to public schools, commercial movie theaters, open fields, museums, universities, fairs, churches, social and civic group gatherings, and foreign audiences and governments. The USDA understood the potential power of its motion pictures and took precautions to make sure the movies did not present the USDA or the government in a way that it considered harmful or insensitive.

The presentation of African Americans in these motion pictures represented the USDA's understanding of black Americans and communicated it to audiences with the endorsement of the US government. The result, I argue, is a view of African Americans as inferior and a threat to white society, a fundamental precept of American racist ideology that predated the films but was maintained and supported by the motion pictures. Each motion picture presented a carefully crafted filmic reality designed to communicate to the USDA audience messages that were fundamentally influenced by the long-standing racist ideology that defined blacks as inferior to whites and therefore in need of control and supervision. This racist belief maintained that without white control and supervision, blacks threatened white society in a variety of ways, including social and economic. In order to demonstrate this conclusion, I address three fundamental themes of the USDA's filmic representation of African Americans that bridge the films and constitute the major findings of these analyses: (1) segregation is sacrosanct, (2) southern African Americans are farmers, and (3) blacks are a problem population.

Segregation Is Sacrosanct

The overriding ideological theme of all the USDA motion pictures about African Americans made from 1921 to 1942 is that segregation is sacrosanct. At no time in any of the films was segregation challenged; thus the USDA films conveyed the idea of segregation as the proper condition for African Americans in the United States. The fact that segregation denied African Americans their civil rights did not inhibit the USDA from supporting segregation in its films without question or comment. The USDA's filmic protection of segregation may be more fully demonstrated by examining how the motion pictures supported the doctrine of separate but equal, argued that Jim Crow laws were not harmful, used stereotypes to demean African Americans, and suggested that there is equal opportunity in segregation. By utilizing these four subthemes, the USDA motion pictures argued that segregation is sacrosanct and unquestionably the proper state for African Americans.

Separate but Equal

Author Jerrold Packard explains that the Supreme Court's 1896 ruling in *Plessy v. Ferguson* meant that discrimination was deemed legal when equal services or facilities were provided for blacks. The result was that the "'separate but equal' provision would quickly grow into the legal framework by which Jim Crow laws would . . . be written and enforced."[2] Of course, there was nothing equal in the doctrine of separate but equal.

It existed as a means for upholding segregation without providing truly equal opportunities, services, or facilities for African Americans. The USDA's motion pictures supported the separate-but-equal credo by maintaining the separation of whites and blacks both visually and narratively.

Visually, the films maintained the separation of whites and blacks by avoiding scenes in which blacks and whites were seen together and by showing whites as authority figures such as USDA officials, extension agents, supervisors, landlords, doctors, nurses, or program leaders when interacting with blacks. The motion pictures did not include scenes of whites and blacks together in social venues, workplaces, or public venues such as schools and grocery stores. The most extreme example of the relegation of blacks to a black world is *Henry Browne,* where the Browne family existed in a segregated world both on their farm and when visiting their son serving with the all-black Tuskegee Airmen. However, all of the films visually segregated blacks and whites and thus visually communicated to audiences that separation was the unquestioned condition for blacks and whites in the United States. Blacks were shown in segregated schools, social venues, 4-H clubs, churches, grocery stores, and syphilis treatment programs and in an all-black military unit. The force of these depictions reinforced a view of African Americans as having separate-but-equal facilities, opportunities, and services despite the harsh reality of Jim Crow racism.

The USDA motion pictures maintained the doctrine of separate but equal by emphasizing the distinct differences between the facilities and services offered to blacks. In *Helping Negroes* and *The Negro Farmer,* the black extension service and its agents were touted as the service provided for African American farmers. Also, in these films, Tuskegee Institute was lauded as a top-notch school for the education of blacks. In *Three Counties* the clinics were clearly publicized to the black community as a public health service provided for African Americans. Thus, in the USDA's motion pictures, the doctrine of separate but equal was upheld narratively by the USDA. For example, in *Three Counties* the film asked, how will poor blacks living in the piney woods receive syphilis treatment? The film explicitly answered the question by sending mobile clinics, facilities set up for the detection and treatment of syphilis, to the backwoods communities. *Negro Farmer* wondered who could help these wretchedly poor black farmers living on depleted soil in unhealthy conditions. The film's narrative demonstrated that the black extension service was the USDA's answer, including black 4-H clubs for African American youth.

Thus, separate but equal was always supported in the USDA motion pictures, both visually and narratively. Segregation was not questioned or negatively commented upon. The result was that the USDA communicated a clear message to the audience: segregation works for blacks in the United States, and the USDA services helped blacks improve their lives within the boundaries of segregation through the false assurances of the doctrine of separate but equal.

Jim Crow

Separate but equal was the provision that allowed Jim Crow laws to stand legally. But Jim Crow was more than just legal statutes; it was also a way of life in the South. Scholar Jane Rhodes explains, "Jim Crow racism was reinforced by a system of economic exploitation, political disenfranchisement, and violence targeted at blacks."[3] The threat of racial violence to blacks included beatings, torture, and lynching. Packard explains that Jim Crow equality "meant that the Negro part of everything would be indisputably and often grotesquely inferior to its white counterpart."[4] The harsh realities of Jim Crow were many, including African Americans being denied jobs, the right to vote, and equal access to education and other public services provided to whites. Hence Jim Crow racism was devastating to southern blacks, who had no recourse or way to combat its pervasiveness.

USDA officials were well aware of the results of African American disenfranchisement, including racial violence. Additionally, the Booker T. Washington essay "The Case of the Negro," partially quoted at the end of *Negro Farmer*, includes passages, not mentioned in the film, arguing against denying blacks their civil rights. Yet the USDA films ignored the injustice of Jim Crow laws and customs in the South. Instead, they presented the problems facing blacks as the result of the practices of black farmers (such as the overuse of land to raise cotton as a cash crop), a position that fundamentally ignored the social, political, and economic reality of Jim Crow in the South.

The problems faced by African Americans in these motion pictures (substandard housing, unsanitary living conditions, the brutal tenant farming system, lack of nutritious food, and disease) were clearly shown in *Helping Negroes*, *Negro Farmer*, and *Three Counties*, but their root causes, including Jim Crow, were not addressed. By presenting these situations in this manner and by failing to address the brutal oppression of discrimination, the films rhetorically communicated that these problems were the result of actions or inactions of the African Americans represented and suggested that the oppressive Jim Crow system was not the cause.

Granted, it would have been difficult for the USDA to release a film that openly denounced Jim Crow, but by presenting the problems faced by southern blacks as the result of their own misfortune, the USDA implicitly argued that it was the farmers themselves, not the system, that placed them in these dire conditions. If the system was not at fault, as the films argued, then the audience was left to believe that the system was working for southern blacks. Thus, the motion pictures shared and repeated a common strategy: first they demonstrated the conditions, sometimes dire, that African Americans endured in the rural South; they then suggested that the problems resulted from black ineptitude, rather than the oppressive Jim Crow system; and, finally, they suggested that segregated programs could alleviate the problems. For example, *Negro Farmer* suggested that blacks lived on meager farms as a result of poor farming methods that could be improved by African Americans utilizing the techniques taught by black extension agents. *Three Counties* argued that African Americans suffered from the "bad blood," a euphemism for syphilis, and could be cured by the Public Health Service's mobile clinic that operated in the rural black communities. *Henry Browne* suggested that there was no problem; the Browne family thrived within the Jim Crow system and contributed admirably to the US war effort. The films implicitly communicated that Jim Crow was not holding African Americans in economic peonage, that it was not harmful, and that blacks could succeed in the segregated South.

Stereotypes Demean African Americans

Media scholars have successfully demonstrated the destructive nature of racial stereotypes, including their ridicule and contempt of African Americans, their support of white superiority, and their ability to divert attention from black accomplishments.[5] Likewise, the USDA films' stereotypes demeaned African Americans, showing them as second-rate and relegated to inferior positions. Film scholar Donald Bogle explains that some of the earliest black stereotypes were designed to show blacks' satisfaction with their inferior place in American society.[6] The overt and explicit stereotypes used in *Helping Negroes* and the inferential racist stereotypes used in *Negro Farmer* and *Three Counties* represented African Americans as inferior and argued that segregation was needed to separate blacks from whites so that whites could supervise, control, and help blacks who were satisfied with their place in the segregated South.

With the exception of *Henry Browne*, the USDA films portrayed African Americans as inferior to whites through classic stereotypes such as the coon and the pickaninny and through representations of blacks as unintelligent, lazy, unclean, diseased, and unable to enact positive change in their communities without white intervention.

One result of this portrayal was that it diverted attention away from the accomplishments of African Americans. The important pioneering agricultural work done at Tuskegee Institute was mentioned in both *Helping Negroes* and *Negro Farmer*, but the stereotypical representation of African Americans as inferior to whites undermined the importance of these accomplishments and left a lasting impression that blacks needed white supervision and control, a fundamental argument for segregation. Showing blacks living in squalid conditions, riddled with disease, and eating a pellagrous diet while suggesting that white intervention could dramatically improve their lives rhetorically bolstered the concept of white superiority and rationalized segregation to USDA audiences by demeaning African Americans.

Equal Opportunity

In the USDA films some African Americans were presented as successful and able to prosper within segregation, thus suggesting that equal opportunity existed in southern segregation. The three farming films all contained examples of prosperous African American farmers. In *Helping Negroes*, the 1921 film ended by explaining that farmer Rube Collins "becomes a contented farm owner, thankful for the prosperity extension work has helped to bring him." Twenty years later, the same message was communicated in *Henry Browne*, in which the Browne family is shown to be very successful farmers with good crops, a nutritious garden, and healthy livestock while existing in the segregated Jim Crow world. In 1938 *Negro Farmer* showed black farmers who were successful in raising cotton, tobacco, chickens, dairy cattle, and eggplants. *Negro Farmer* also used excerpts from Booker T. Washington's "The Case of the Negro" to argue that African Americans were best able to prosper on the farm.[7] Therefore, the farming films presented successful African American farmers as proof that blacks could prosper in the Jim Crow South.

By showing that African Americans could prosper in the segregated South, the films rhetorically supported the false equality statement in the doctrine of separate but equal. The truth of segregation was inequality for African Americans, who were severely disadvantaged by the system. By presenting prosperous black farmers, the USDA motion pictures communicated the false message that separate but equal worked and did not disadvantage black farmers. It is important that the films communicated that blacks were successful within the constraints of separate but equal because "whites defined the 'proper' place for blacks as being not simply in the South but on Southern farms, well-supervised and uplifted by contact with paternalistic whites."[8] This white paternalism was itself a racist construction; Conrad reports that "the attitude of planters toward their tenants . . . has been called 'paternalistic' but it is hard to imagine a father exploiting his children the way a planter did his tenants."[9] Hence, the racist belief was that blacks benefitted from the system of separate but equal because the supervision provided by white authority figures helped African Americans to be successful southern farmers.

Summary

The USDA films repeatedly communicated that segregation was the proper condition for African Americans by supporting the doctrine of separate but equal, arguing that Jim Crow did not harm southern African Americans, using stereotypes to demean African Americans, and suggesting that the separate-but-equal provision provided equal opportunity to blacks. Through these means, the USDA motion pictures, as a group, communicated the USDA's support for segregation. Conversely, at no time in the films, was it suggested that segregation was one of the "train of evils" that adversely affected African Americans living in the rural South. Through law and custom, Jim Crow solidified relations between blacks and whites. The USDA motion pictures both recognized the harsh reality of Jim Crow and presented segregation as a natural and desired condition for African Americans. Therefore, the USDA films approved segregation as right and just. In short, segregation was the sacrosanct reality that the USDA communicated via its motion pictures.

Southern African Americans Are Farmers

The second pervasive theme presented in the USDA films is that southern African Americans are farmers. The USDA films situated southern African Americans as farmers, demonstrated how they could become better farmers, and highlighted their accomplishments as farmers. The USDA's position was established by arguing that southern blacks belonged on the farm and should remain on the farm and that they were contented and successful in agriculture. The USDA films communicated that farming was the proper life for blacks. However, defining blacks as southern farmers grew from the racist belief that blacks needed to be restricted to the southern farm, where they could easily be supervised by whites in the tradition of the plantation system. Conrad makes this connection when he writes: "Southern tenancy, like slavery before the Civil War, was an institution 'peculiar' to the South. It was not only an economic but a social and political order, the origins of which lay clearly in slavery and the plantation system."[10] Sociologist Martha Myers explains that whites believed that blacks leaving the rural areas to concentrate in cities "threatened public safety" because blacks belonged on southern farms under white supervision.[11] Thus, racist ideology argued for blacks to remain on the southern farm, where they could be supervised by whites as was the practice during slavery. Accordingly, the USDA films argued that blacks belonged and should remain on the southern farm, a position that implicitly argued against black migration away from the southern tenant farming system.

Belonging on the Farm

It is not surprising that the government branch devoted to agriculture in the United States made movies suggesting Americans should be farmers. However, the USDA films analyzed in this book went beyond just suggesting that farming was a suitable pursuit for southern African Americans. The motion pictures began from the a priori position that southern blacks did indeed *belong* on the farm. The films presented African Americans doing other types of work; most notably, they briefly showed black doctors, nurses, USDA extension agents, and ministers at work in the black community. However, these depictions were seen only in their relation to rural southern blacks who were engaged in agricultural and related endeavors. African Americans who worked outside the farm were represented, in a sense, as a support system for black farmers (the implied proper condition for African Americans).

The vast majority of the blacks in the films worked in agriculture, broadly defined. In these motion pictures, African Americans raised cotton, corn, tobacco, poultry, cattle, hogs, and various other agricultural products. They were presented as living on farms with their families. The farms ranged widely in condition and success from shacks in dirt clearings on sour soil to Henry Browne's model forty-acre farm. But even when considering the worst farms on the most meager soil, the films suggested that black families should live on the farm and look for ways to improve their farming. No matter how poor the families were or how impossible their situations, the movies never suggested that the occupants would be better off anywhere else but on the farm.

Two of the films, *Helping Negroes* and *Negro Farmer*, were constructed specifically to suggest that the USDA black extension program could help any black farm family make a better life in agriculture. The results were often miraculous as the views changed from shanties on ill-kept farms to modest homes on successful farms. In *Helping Negroes*, the Collinses were reported to be contented farmers after learning how to make the most of their situation. Twenty years later, the USDA's portrayal of the successful African American farm family on a well-managed and productive farm, *Henry Browne*, was the cinematic pinnacle of the department's position that southern blacks belonged on the farm.

The films began and ended with African Americans living on farms, and no other option was ever explored. The fact that southern African Americans should be or were engaged in other pursuits was never seriously broached. Although never explicitly stated, the chronology was clear: African American slaves worked in southern agriculture, black tenant farmers were engaged in southern agriculture, and, even during World War II, the proper place for southern black families was the farm in the segregated South.

Situating southern African Americans on the farm was important to the USDA's presentation of their work. The USDA touted data proving that several million southern African Americans were engaged in agricultural work, and they felt some pressure from external groups to improve the opportunities for southern blacks. The department used this as the basis for the creation of its cinematic treatments of African Americans and faithfully communicated this message to its audiences.

Remaining on the Farm

The USDA farming films sent a consistent message that despite all the problems southern blacks confronted, there was no sufficiently compelling reason for African Americans to leave the farm for other pursuits.

For example, when faced by the scourge of the boll weevil, tenant farmer Rube Collins was helped immeasurably by the efforts of the extension service in *Helping Negroes.* By the end of the film the Collinses are presented as contented farm owners, their farm having been transformed without further mention of the boll weevil. *Henry Browne* introduced the audience to the Browne family and their model forty-acre farm in Alabama during World War II. The fact that the Browne family farm was successful in every respect was good for the Brownes and good for America as they did their patriotic duty both on and off the farm. Both films communicated that these black families should not leave their farms, as they were successful, happy, and productive.

The strongest argument for remaining on the southern farm was found in *Negro Farmer.* This 1938 USDA motion picture begins by introducing a 110-year-old black woman, Aunt Sally Smith, who was born in Africa and, undoubtedly, brought to the United States as a slave. The narrator explained that she has "lived to see the hard lot endured by her generation and that of her children in some measure bettered by [the USDA] campaign to help Negroes help themselves." Thus, the film linked the agricultural realities of the 1938 southern black farmer to the legacy of slavery in the South. From there the movie addresses the grim conditions that most southern black farmers endured: "All too many barely make a living. In common with all southern farmers they have suffered a train of evils that have their roots in poor farming, particularly one-crop farming." *Negro Farmer* set as its foundation the harsh realities of black tenant farming by explaining that it was a hard lot and difficult to make a living. The remainder of the film, as I discussed in Chapter 3, presents a series of improvements that made life on the farm better, although it was not always clear how desperately poor farmers could avail themselves of these options. Thus, *Negro Farmer* also communicated the USDA's resolute message that blacks should remain on the southern farm, no matter their material conditions, because success was always just one scene away.

It was the end of the motion picture that made the most striking rhetorical argument for blacks remaining on the southern farm. After showing some of the most starkly desperate scenes of southern black farm life in USDA motion pictures, the movie ends with a sequence devoted to specific lines from Booker T. Washington's 1899 article "The Case of the Negro" to argue that blacks should remain on the southern farm. The narrator quotes Washington: "Since the bulk of our people already have a foundation in agriculture, [they] are at their best living in the country engaged in agricultural pursuits."[12] Hence, the USDA made its argument explicit by borrowing from Washington's essay published nearly forty years earlier: African Americans should remain on the southern farm. The thrust of this sequence was an implied argument against the great migration to the north; the USDA was well aware that the migration was not supported by southern whites who relied on black labor and believed that the proper place for blacks was on the southern farm.

The USDA motion pictures explicitly and implicitly argued against blacks leaving southern farms for other pursuits. The films never mentioned nonagricultural issues that might provide valid ways to improve the lives of blacks living on southern farms, such as those mentioned by black extension agent Thomas M. Campbell in his report to the USDA after visiting the North to see "what is happening to the colored people who have migrated."[13] The 1924 USDA internal memo stated that as a result of his investigation Campbell believed that blacks who migrated north found "larger opportunity for development, better wages, less restrictions, and better school facilities."[14] The USDA could have suggested ways for improving the lives of southern blacks, such as by offering better educational opportunities, but instead resolutely maintained its message that blacks should remain on the farm and improve their agricultural endeavors.

Despite having its own report suggesting that migrating blacks found better conditions in the North, fourteen years later the USDA made *Negro Farmer,* which remained the USDA's most explicit cinematic argument that blacks should remain on the southern farm working in agriculture, and that message was faithfully communicated in all of the films analyzed in this book. By holding to this position, the USDA communicated its message that blacks should remain on the southern farm despite the formidable challenges black tenant farmers faced in the Jim Crow South.

Successful and Contented in Southern Agriculture

The third part of the USDA's rhetorical argument centered on the position that African Americans were successful and contented southern farmers. Each of the USDA farming films explicitly argued that African Americans were effective and happy farmers or could be with a little help from whites. *Helping Negroes* ends by showing how conditions on the Collins farm dramatically improved and presents the Collins family as gratified farmers. Despite the problems the family faced, through USDA demonstration work the family's challenges were eliminated and they became happy, successful farmers.

Despite showing some of the most desperately poor black farm families in Alabama, *Negro Farmer* focused much attention on successful black farms that raised various crops and livestock, even a thriving commercial eggplant farm. The USDA's argument that blacks were successful farmers was most overtly stated in the Booker T. Washington sequence ending *Negro Farmer.* Paraphrasing Washington's essay, the narrator says, "Dr. Booker T. Washington taught that agriculture should be the fundamental pursuit of his race in America. The Negro, he said, must begin at the bottom and lay a sure foundation, not be lured by any temptation to rise on a false foundation. Progress by any other method will be but temporary and superficial." The film purported that the only sure foundation for African American success was agriculture. Thus, the film argued that farming was the true vocation for African American farmers and showed that African Americans are successful on the southern farm.

Henry Browne was the USDA's pinnacle film of prosperous, satisfied black farmers. In this motion picture the audience met the Brownes, the perfect successful black farm family. Henry Browne was the impeccable farmer; he made the most of his forty-acre farm and succeeded in all of his endeavors. From the USDA's perspective, Browne and his family were the quintessential efficacious black farmers and patriots because they raised good crops, took care of the land and equipment, fed themselves, and gave their oldest son for military service. The Brownes were presented as happy, contented, and successful in their endeavors.

In these films the audience was presented with comfortable and effective black farm families who had much to show for their hard work and many reasons to be proud of their success. All three of the farming films explicitly communicated that African Americans could be and were efficacious and satisfied in southern agriculture. Therefore, success on the southern farm became the third part of the USDA filmic message that supported both the a priori position that blacks were meant to be farmers and the subsequent argument that African Americans should remain on the farm. Contented success becomes the linchpin argument: blacks belong on the farm, are successful as farmers, and, therefore, should remain on the farm.

However, many thousands of southern African Americans were not contented in the South. As Isabel Wilkerson explains in her book *The Warmth of Other Suns,* the reality for southern blacks was that "their every step was controlled by the meticulous laws of Jim Crow. . . . It afflicted the lives of at least four generations and would not die without bloodshed, as the people who left the South foresaw."[15] African Americans fled the South to find better lives elsewhere. They left by the millions because they believed a better life awaited them beyond the racist reality of the Jim Crow South. The USDA films avoided addressing both the African Americans' desire to leave the South and the reality that many were going north every year. Instead, the USDA represented southern blacks as contented and successful farmers and ignored the reality of the ravages of Jim Crow on the lives of southern blacks. The contentedly successful Collins and Browne families denied the need and the desire for migrating north. Denying that reality supported the racist belief that blacks belonged on southern farms, where they could be controlled. As Wilkerson explains, "The Migration did not end until the 1970s, when the South began finally to change—the whites-only signs came down, the all-white schools opened up, and everyone could vote. By then nearly half of all black Americans—some forty-seven percent—would be living outside the South, compared to ten percent when the Migration began."[16] However, for decades the USDA represented southern blacks as contented and successful farmers with no reason or need to leave their farms or the South.

Summary

The implicit and explicit arguments that southern blacks belonged on the farm, should remain on the farm, and were content and successful in agriculture constitute the second theme of the USDA films: African Americans are farmers. This stance allowed the USDA to define the problems faced by African Americans as farming challenges and to offer agricultural solutions. Thus, the positive changes that occurred in the lives of black farmers were ascribed to good farming practices taught by the USDA. More significantly, however, the argument that blacks were meant to be farmers is relatable to the racist belief that blacks threatened white society by leaving southern farms, where they could be supervised by whites, and migrating into urban areas, where their unrestricted contact with whites was seen as threatening.

A Problem Population

The third persistent theme demonstrated in the USDA motion pictures is that African Americans were a problem population. The motion pictures never overtly stated that blacks were a problem population; instead they strongly implied it as they demonstrated time and again that African Americans required white authority and help, white control and intervention, and separation from white society. The USDA motion pictures argued that African Americans were fundamentally different from whites and that this difference meant that blacks were a problem population.

Fundamentally Different

The fundamental belief that African Americans were different from white Americans was what drove the USDA filmmakers to produce these films. The essential position of the USDA was that because of this difference, special films were needed to deal with the topic of southern black farmers.

The first two films made this fact clear in their titles: *Helping Negroes to Become Better Farmers and Homemakers* and *The Negro Farmer.* The titles explained that the films were about farmers who were not white and therefore different from white farmers. The premise of these films was that blacks had special problems that required special assistance from the USDA. In *Helping Negroes* the unstated question asked, how can the USDA help black farmers? The answer the film provided was to offer a black extension service that adhered to the segregated Jim Crow reality of the southern states. Likewise, in *Negro Farmer* the answer was once again to utilize black extension in its segregated context but also to demonstrate how segregated 4-H clubs could teach younger African Americans how to be better farmers (within the segregated system). In these films the difference was clear: non-African American farmers were farmers, but African American farmers were a special problem population. Thus, African Americans were defined as fundamentally different from whites, and that difference was shown as a problem that had to be addressed by the USDA.

In *Three Counties* the distinction between blacks and whites was not explicitly stated in the title, and, at least in the early portions, the movie suggested that the syphilis testing and treatment program was for whites and blacks. However, the motion picture quickly abandoned that position to dwell entirely on the black community. For example, flyers were distributed that read "Colored People, Free Blood Tests, Free Treatments, Come and bring all your family." An entire sequence was devoted to a black minister addressing his African American congregation, the audience was taken to an all-black school and told that all the children had been tested, and viewers visited a dance hall where young black men and women were tested. Hence, *Three Counties* was specifically about African Americans and the special efforts that the Public Health Service took to reach black communities for testing and treatment.

This subsequent and abrupt switch to a singular emphasis on African Americans highlighted the fact that the program and the movie defined blacks as different from whites and in need of special treatment; for example, syphilis was called the "bad blood" only when referring to communications about the disease with African Americans.[17] *Three Counties* demonstrated that the syphilis testing and treatment program grew from the racist ideological perspective of racial medicine holding that African Americans were physically different from whites and that poor black health was a problem for white society.

Henry Browne took the most extreme position in favor of black difference by highlighting the racist reality of segregation in the South. This movie presented Henry Browne and his family as completely segregated from white society. Although successful—indeed, a model for small farmers—the Browne family appeared to have no meaningful interaction with whites either on the farm or when they visited their son training with the all-black 99th Pursuit Squadron, the Tuskegee Airmen. The result was that blacks were seen as different and separate from whites in a cinematic version of southern segregation that left one NAACP officer hoping that "other films will follow including those which will show Negroes and whites working together as fellow Americans."[18] *Henry Browne* highlighted the USDA's perspective that blacks were different from whites by presenting African Americans in a totally segregated world. Ideologically, *Henry Browne, Farmer* used this rhetorical device in order to communicate its message.

Therefore, each film, in its own way, demonstrated the USDA's position that African Americans were different from whites and showed how blacks should be treated differently as a result. The films suggested special programs, special strategies, and special circumstances for African Americans. Regardless of the individual filmic situations, when combined, the movies persuasively argued that blacks were different from whites and required special treatment because their difference presented them as a problem population.

Defining African Americans as different from white Americans was a necessary step before arguing that blacks were a problem population. Unfortunately, three of the USDA films, *Helping Negroes, Negro Farmer,* and *Three Counties*, explicitly presented blacks as a problem. The fourth film, *Henry Browne*, did not overtly argue that African Americans were a problem population but did argue that blacks and whites must remain separate. Together the four films provided a broad view of blacks as a problem group that needed to be treated differently and separately from whites.

Summary

By demonstrating that African Americans were different from whites and needed special attention and treatment, the USDA's motion pictures argued that African Americans were a problem population. Sociologist Martha Myers explains that southern whites perceived blacks as a problem population because they were a threat to white hegemony and black labor was a necessary resource for many whites.[19] Thus, as Stuart Hall explains, blacks were ideologically constructed as a "problem population."[20] The USDA films, supporting this ideological construction, rhetorically argued that African Americans were different and required special treatment to protect white society. Thus, blacks were shown as being in need of white authority and control, and this message was communicated to the USDA audiences.

In the preceding sections I argued that combining my analyses of the four films discussed in this book sheds light on the USDA's officially sanctioned representation of African Americans. What is revealed is a view of African Americans that is decidedly prejudiced by racist ideologies. Taken as a whole, these films indicate that the USDA held that segregation was sacrosanct, African Americans were farmers, and blacks were a problem population. These three themes work together to communicate the racist belief that blacks were inferior and threatened white superiority.

The USDA Films and a Threatening Black Inferiority

The three themes discussed above—that segregation was the correct condition for blacks, that blacks were farmers, and that African Americans were a problem population—were evident because the USDA motion pictures dedicated to African Americans in the rural South supported the racist ideologies that defined blacks as inferior and a threat to white society. Both *Helping Negroes* and *Negro Farmer* took as their starting point the difference between blacks and whites. Once establishing this, the motion pictures used overt and inferential racist stereotypes to demonstrate that blacks were inferior to whites. Film scholar Donald Bogle argues that the earliest racist stereotypes in films were used "to entertain by stressing Negro inferiority,"[21] and media scholar Jane Rhodes writes that from its earliest years, American "cultural discourse reinforced the inferiority of blacks with highly visible stereotypes."[22] The inferiority of blacks was central to arguments in favor of slavery, white supremacy, and segregation, the racist argument being that, since blacks were inferior, they needed white control in order to exist in white civilization. Without white control, the racist ideology argued, African Americans threatened white society with their barbaric tendencies.

Three Counties presented the argument that blacks were inferior and a threat to white society via the racist beliefs of racial medicine. Racial medicine saw blacks as physically and morally inferior to whites. Therefore, practitioners of racial medicine believed that blacks had higher rates of certain diseases, such as syphilis. In *Three Counties*, blacks were the source of contagious disease because they were racially inferior, the foundational belief of racial medicine. And because African Americans interacted with whites, within the limits allowed by segregation, black disease threatened white health while black immorality threatened white society.

Henry Browne avoided presenting racist stereotypes and did not deal explicitly with African American health beyond nutrition. However, the basis of segregation was the racist belief that blacks were different, inferior, and needed to be separate from whites. Federal Appeals Court judge and author A. Leon Higginbotham Jr. explains this belief as the "precept of inferiority." Higginbotham explains that "our nation was founded explicitly, prospered implicitly, and still lives uneasily on the precept of black inferiority and white superiority. Indeed, that precept helped to legitimize slavery in America and served to justify segregation of African Americans in this nation long after slavery had been abolished. To this day, the premise of black inferiority and white superiority remains an essential element of the 'American identity,' mesmerized as we still are by race and color."[23] Segregation due to inferiority resulted from the belief that blacks threatened white society in a variety of ways. Author Jerrold Packard argues that an important impetus of Jim Crow laws was that whites "grew more physically fearful of blacks" in the early twentieth century as some blacks "began to press for greater opportunities."[24] Likewise, sociologist Martha A. Myers explains that the concept of black threat helps to explain why whites saw blacks as a menace, as "blacks threatened the dominant position of whites in the political and social hierarchy."[25] Therefore, whites sought ways to control blacks via legislation, custom, and violent coercion. Moreover, Higginbotham contends that segregated education kept blacks and whites separate because integrated education would allow interbreeding, which was considered a threat to white society.[26] This perceived black threat of African American inferiority drove whites to enforce segregation to protect white superiority. Hence, all four films, in their own way, enacted black inferiority and implied a black threat to white society. The result was that the films reinforced fundamental elements of the racist ideology that blacks were inferior to whites and a threat to white society.

The USDA films were part and parcel to the complex governmental support of segregation in the United States that allowed Jim Crow disenfranchisement of African Americans, and its films reflected this reality. As Rhodes explains, the media communicate the "beliefs and values" of a racist society.[27] Moreover, Hall explicitly connects the state and the media through basic social relations: "Whereas the media, like the state, have a structure, a set of practices which are not reducible to the individuals who staff them. What defines how the media function is the result of a set of complex, often contradictory, social relations."[28] Given that the doctrine of separate but equal defined social relations and governmental policies, the USDA films reflected the harsh reality of Jim Crow while presenting segregation as a natural and desired condition for African Americans. Therefore, the USDA films approved segregation as right and just and, in so doing, perpetuated and bolstered the belief in a threatening black inferiority.

During the period these films were made, as well as before and after, blacks were migrating away from the southern farms, often to the North in search of better conditions, but also to urban areas in the South. Myers explains that blacks leaving the rural areas to concentrate in cities concerned whites, who felt that this situation "threatened public safety" because blacks who moved to the urban areas lacked white supervision.[29] This fear of blacks concentrating in urban areas and the belief that blacks belonged on southern farms are fundamental aspects of the racist ideology in which the "inferiority of African Americans was given the standing of a natural principle embodied through the moral and social climate of the time that was not to be questioned."[30] Therefore, the USDA films reinforced the position that blacks should remain on southern farms as farmers, where they were more easily controlled and posed less of a threat to white society.

Likewise, American racist ideology, as constructed in the southern United States in the early twentieth century, was fundamental to the USDA's official institutional understanding of African Americans as a problem population. After the abolishment of slavery and the end of the period of Reconstruction, southern whites feared that free blacks threatened white society. Judge Higginbotham says that a fundamental aspect of racist belief was that once the "restraints of slavery" were removed, "blacks had to be segregated from whites because, left entirely to their own devices, free blacks would tend to corrupt the moral virtue and physical purity of white society."[31] White superiority saw blacks as inferior and a threat to white society in many ways; most fundamental was a fear that interracial sex would corrupt white blood.[32] But the racist belief in black inferiority maintained that blacks were inferior in every way, including morally and physically, and their existence threatened white health, purity, society, and hegemony. Whites created Jim Crow laws to separate blacks from whites and maintain white hegemony by imposing separate, but unequal, services and facilities. After Reconstruction, blacks existed as a disenfranchised group "trapped in economic peonage."[33] Packard argues that "concerns of Northern abolitionists had largely evaporated by the 1890s, and the entire nation seemed content to regard African Americans precisely as Southern racists urged: as a hopelessly inferior underclass, whose direction was best left to the South's lead."[34] With its motion pictures, the USDA did not locate the problems of the black community within the history of slavery, segregation, and economic despair; instead it divorced blacks from their historical situation and presented their problems as a result of their "natural" inferiority.

By presenting African Americans as inferior and in need of segregation and white control and supervision, the USDA films reflected, supported, and reinforced American racist ideologies based upon white superiority and black inferiority that fueled slavery, segregation, and Jim Crow racism. They also served a practical propagandistic role for the USDA, as they variously touted the department's programs and suggested to its critics that the USDA was taking action to help black farmers. Moreover, the films demonstrated how racist ideology shaped the USDA's perception of African Americans and dictated its filmic representations of blacks to audiences for decades.

A Critical Media Literacy

I began this book by mentioning that scholars have long been interested in the ways that the media communicate racist ideologies. This book is an in-depth study of how USDA motion pictures focusing on African Americans made from 1921 to 1942 communicated messages that maintained and bolstered racist ideologies. Thus, this study adds to the body of film scholarship that exposes the role of the media in producing and reproducing positions that for too long have contributed to racist beliefs and values. In fact, racist beliefs continue to be perpetuated in the media today.[35] Therefore, it is important that media scholars not only continue to expose these problems but also suggest ways to combat racist worldviews in the media. Using scholarship such as the current study as part of critical media literacy is one way to help individuals understand how the media communicate racist ideologies. Therefore, it seems fitting to end this book with a brief discussion of critical media literacy as part of critical media pedagogy.

One important goal of critical media pedagogy is to aid efforts to bring about positive social change. Stuart Hall argues this explicitly when he urges, "We must attend to the complexities of the ways in which race and racism are constructed in the media in order to bring about change."[36] The current study demonstrates how racist ideologies were constructed and communicated through USDA motion pictures and, in so doing, shows how a government institution contributed to the oppression of African Americans by supporting the precept of inferiority.[37] Although the USDA filmic texts are matters of history, similar problems persist in the twenty-first century. Racist ideologies are resilient and continue to thrive, and the media continue to reproduce and reinforce them in a variety of ways both overt and implicit. By studying how racist ideologies affected specific media productions in the past, we can better learn how to consume, interrogate, and interpret current media and use that understanding to encourage change. The change may be helping media producers understand how to avoid reproducing racist ideologies, or it may be part of a larger critical media pedagogy designed to help media consumers to recognize, interrogate, and resist ideological suppositions in mediated texts.

Media scholar Douglas Kellner argues that media studies that highlight race, gender, and class issues can be "part of a critical media pedagogy that . . . can empower people to gain sovereignty over their culture and enable them to struggle for alternative cultures and political change."[38] The African Americans presented in the USDA films had no ability to challenge or shape their representations in the films. The subjects of the films were never allowed to address their situations. The films were carefully constructed to present specific viewpoints and communicate crafted messages from the USDA's institutional view, not from the viewpoint of the black farmers. As a result, the filmic messages maintained and reinforced the racist ideologies of the time. The critical media pedagogy that Kellner and others envision could give individuals the knowledge and skills they need to resist harmful dominant perspectives and create messages that work to aid social and political change.

One part of that pedagogy focuses on helping individuals understand the media. We find our lives increasingly engulfed by media texts. Moreover, the traditional boundaries between media are breaking down at an astonishing rate. Traditional print media such as books, magazines, newspapers, and textbooks are consumed in electronic formats on devices that allow both planned and unplanned multimedia opportunities. From a single handheld device with wireless capability, a reader can go from an electronic version of a print book to filmic adaptations of the work, songs inspired by the work, and interpretations of the work by professional reviewers and amateur bloggers. Concurrently, for those with Internet connectivity, access to media texts is growing exponentially. Television programs, films, newspapers, radio shows, books, and a seemingly endless amount of content on the Internet are available instantaneously to consumers of almost any age. This unprecedented access is a boon for people seeking information, education, and entertainment, but it also creates an environment where the differences between media formats, genres, and credibility are blurred. The consumer may or may not be equipped to readily discern the reliability of the message being communicated or have the skills to challenge or question the viewpoint presented. Therefore, now more than ever before, people need to be able to understand, question, challenge, and evaluate media for themselves. One way to address this need is through critical media literacy.

Media scholars Douglas Kellner and Jeff Share declare, "Media literacy helps people to use media intelligently, to discriminate and evaluate media content, to critically dissect media forms, to investigate media effects and uses, and to construct alternative media."[39] Critical media literacy can provide individuals with the ability to discern the differences between the news coverage of an event and a blogger's opinion of the event. Moreover, media literacy can also show how institutionalized messages can be affected by ideologies that communicate racist views of the world.

Kellner and Share argue, "There is expanding recognition that media representations help construct our images and understanding of the world and that education must meet the dual challenges of teaching media literacy in a multicultural society and sensitizing students and the public to the inequalities and injustices of a society based on gender, race, and class inequalities and discrimination."[40] This is a worthy goal, as Stuart Hall warns that "inferential racism is more widespread [than overt racism]—and in many ways, more insidious, because it is largely invisible even to those who formulate the world in its terms."[41] The present study demonstrates how racist ideologies were inscribed in the USDA motion pictures and communicated to audiences in both overt and inferential ways. In this way the study might be used to suggest how current officially sanctioned productions may also be affected by racist ideologies. This type of in-depth study, dealing with media texts from the past, can be used to educate students about topics such as overt and inferential racism and to demonstrate how racist ideologies are reproduced in media texts. Thus, the current study can play a role in sensitizing students to these issues so that they may be better able to spot similar problems in other mediated texts.

Further, Douglas Kellner argues, "A media literate person is skillful in analyzing media codes and conventions, able to criticize media stereotypes, values, and ideologies, and thus literate in reading media critically."[42] The current study elucidates how the USDA films used stereotypes that denigrated African Americans and exposed the racist values and ideologies inscribed in the films. Through textual analysis and the use of archival materials in the USDA archives, the study dissects the filmic texts in order to expose the underlying racist messages. Thus, this study can serve as one of many works that could educate students to critically read and challenge media texts.

Finally, Kellner maintains that "critical media literacy is necessary to develop educated students and citizens, since media culture strongly influences our view of the world."[43] The USDA films examined here presented a view of African Americans that invited audiences to see blacks from a racist worldview steeped in the belief that blacks were inferior to whites in every way and therefore in need of segregation and control. Whether this or similar studies are used, developing and practicing critical media literacy in schools is important. By teaching young people how to be smart consumers of mediated texts, we provide them with the ability to better understand their world.

Notes

Preface

[1] Hall, "Whites of Their Eyes," 35.
[2] Rhodes, "Visibility of Race."
[3] Bogle, *Toms, Coons, Mulattos*; Cripps, *Slow Fade to Black*; Diawara, "Black Spectatorship"; Hall, "Whites of Their Eyes"; Rhodes, "Visibility of Race."
[4] Noble, *Negro in Films.*
[5] Bogle, *Toms, Coons, Mulattos*; Bobo, *Black Women*; Brinson, "Myth of White Superiority"; Hall, "Whites of Their Eyes."
[6] Hall, "Whites of Their Eyes," 41.
[7] Carey, *Communication as Culture*; H. Gray, "Television, Black Americans"; Hall, "Rediscovery of Ideology."
[8] Ross, "Struggles for the Screen."
[9] MacCann, *People's Films.*
[10] Raymond Evans, "The U.S.D.A. Motion Picture Service, 1908–1943," 8.
[11] "The Use of Educational Motion Pictures by the United States Department of Agriculture," September 1929.
[12] MacCann, *People's Films*, 55.
[13] Noble, *Negro in Films*, 106.
[14] Winn, "Documenting Racism."
[15] Note that the film is most consistently called *Helping Negroes to Become Better Farmers and Homemakers* in USDA materials. However, the film is also sometimes titled *Helping Negroes Become Better Farmers and Homemakers* (dropping the word "to"). Moreover, the film is infrequently referenced with other similar titles (often shortened versions).
[16] I would encourage future research into this area, as I believe it would be important scholarship.
[17] Barnouw, *Documentary*, 85.
[18] Ibid.
[19] Ellis and McLane, *New History of Documentary Film*, 1.
[20] Ibid., 1–3.
[21] Ibid., 2.
[22] "Use of Educational Motion Pictures," September 1929; Fred W. Perkins, "A Twelve-Year Trial of Educational Films," 51.
[23] Nichols, Representing Reality, 115.
[24] Hall, "Culture, the Media"; Jameson, *Political Unconscious.*

[25] Aden, "Back to the Garden"; Brinson, "Myth of White Superiority"; Brummett, "Electric Literature"; Frentz and Farrell, "Conversion of America's Consciousness"; Frentz and Hale, "Inferential Model"; Frentz and Rushing, "Rhetoric of *Rocky*"; Payne, "*The Wizard of Oz*," "Characterology," "Political Vertigo"; Rasmussen and Downey, "Dialectical Disorientation in *Agnes*," "Dialectical Disorientation in Vietnam"; Rushing, "Rhetoric," "Mythic Evolution," "Evolution," "Frontierism"; Rushing and Frentz, "Rhetoric of *Rocky*," "*Deer Hunter*," "Frankenstein Myth"; Solomon and McMullen, "*Places in the Heart*"; Winn, *American Dream*.

[26] Blakesley, *Terministic Screen*, 5.

[27] Simpson, "*Copycat*, Serial Murder," 148.

[28] Benson, "Looking for the Public," 130.

[29] Blakesley, *Terministic Screen*, 5.

[30] Note that an earlier version of the research on this motion picture was published in Winn, "Documenting Racism."

[31] For more on the idea of economic peonage, see Packard, *American Nightmare*.

Chapter 1: A Brief History of the USDA Motion Picture Service to 1943

[1] In this book I use the Motion Picture Service (MPS) to refer to the USDA unit that was charged with the production of motion pictures for the Department of Agriculture. This is consistent with the use of the term in USDA and is the term chosen to delineate the motion picture materials in the USDA archives. However, it is important to note that the unit was called a variety of different names during its existence, including the Office of Motion Pictures and the Division of Motion Pictures.

[2] Evans, "U.S.D.A. Motion Picture Service, 1908–1943."

[3] Ellis and Thornborough, *Motion Pictures in Education*, 17.

[4] "U.S. Takes Lead in Educational Film Production," *New York Herald Tribune*, Sunday, April 1, 1928.

[5] Ellis and Thornborough, *Motion Pictures in Education*, 18.

[6] Evans, "U.S.D.A. Motion Picture Service, 1908–1943," 3.

[7] Ibid.

[8] Untitled report, n.d.

[9] Ellis and Thornborough, *Motion Pictures in Education*, 10.

[10] Untitled report, n.d.

[11] Ibid., 17. Interestingly, Clime would later become a noted landscape painter.

[12] Ellis and Thornborough, *Motion Pictures in Education*, 18.

[13] Evans, "U.S.D.A. Motion Picture Service, 1908–1943," 4.

[14] "No. 55, Regarding Motion Picture Activities of the Department," memorandum, December 18, 1913.

[15] "From the Report of the Editor, Department of Agriculture, 1913–1916," n.d.

[16] Ellis and Thornborough, *Motion Pictures in Education*, 19.

[17] "For Mr. G. W. Wharton," memorandum, February 4, 1914.

[18] "For Dr. Galloway," memorandum, February 16, 1914.
[19] "No. 85, The Purchase of Motion-Picture Apparatus," memorandum, April 20, 1914.
[20] C. W. Warburton, "A World Extension Course Conducted through U.S. Films," February 17, 1930, 2.
[21] "To the Secretary of Agriculture on the Work of the Committee on Motion Picture Activities," memorandum, November 30, 1914, 9.
[22] "Motion Picture Policies of the U.S. Department of Agriculture," January 31, 1923.
[23] "Annual Report of the Office of Motion Pictures," June 30, 1923; Ousley to Lemmle, December 5, 1917.
[24] "Informal Report of Motion Picture Work, April 1 to 15, 1918," April 15, 1918.
[25] "Report to the Secretary of Agriculture on the Work of the Committee on Motion Picture Activities," November 30, 1914, 8.
[26] Ball to Huston, April 2, 1913.
[27] "For Heads of Bureaus, Offices, and Independent Divisions," memorandum, May 24, 1913.
[28] "For Dr. Galloway," memorandum, January 6, 1914.
[29] "To the Secretary of Agriculture on the Work of the Committee on Motion Picture Activities," memorandum, November 30, 1914, 2.
[30] Evans, "U.S.D.A. Motion Picture Service, 1908–1943," 4.
[31] Ibid., 5.
[32] Ibid.
[33] "Use of Educational Motion Pictures," September 1929, 2.
[34] Ibid.
[35] "For the Secretary," memorandum, July 7, 1917.
[36] Ousley to Bacon, July 25, 1917.
[37] Ousley to Lemmle, December 5, 1917.
[38] "From the Report of the Editor," 4.
[39] "Informal Report on Motion Picture Work, March 1 to 15," March 8, 1918, 2.
[40] "Informal Report of Motion Picture Work, April and May, 1919," n.d., 2.
[41] Evans, "U.S.D.A. Motion Picture Service, 1908–1943," 5–6.
[42] Ibid., 8.
[43] Chester A. Lindstrom, "Films Working for Agriculture," April 18, 1945, 2.
[44] "Annual Report of the Office of Motion Pictures," June 30, 1919.
[45] "Annual Report of the Office of Motion Pictures," June 30, 1923.
[46] "Use of Educational Motion Pictures," September 1929, 2.
[47] "A Statement of the History and Work of the Motion Picture Section, U.S.D.A.," 2.
[48] "Annual Report of the Office of Motion Pictures," June 30, 1923.
[49] Ibid.
[50] Ibid.
[51] Ibid.
[52] Ibid.
[53] "Annual Report of the Office of Motion Pictures," June 30, 1927.

[54] "Annual Report of the Office of Motion Pictures," June 30, 1928.
[55] "Annual Report of the Office of Motion Pictures," June 30, 1929, 2.
[56] "Annual Report of the Office of Motion Pictures," July 3, 1931, 4.
[57] "Farm Films Seen around the World," newspaper clipping, March 8, 1929.
[58] Ibid.
[59] "Annual Report of the Office of Motion Pictures," June 30, 1929.
[60] "Statement of the History and Work of the Motion Picture Section, U.S.D.A.," 2.
[61] "For Dr. Warburton, Director of Extension Work," memorandum, April 29, 1930.
[62] "Statement of the History and Work of the Motion Picture Section, U.S.D.A.," 2.
[63] "For Dr. Warburton," April 29, 1930.
[64] Ibid., 2.
[65] Ibid.
[66] "Annual Report of the Office of Motion Pictures," July 3, 1931, 5.
[67] Ibid.
[68] Ibid.
[69] Warburton, "World Extension Course," 5.
[70] Ibid.
[71] Warburton to Woods, March 7, 1931
[72] "Annual Report of the Office of Motion Pictures," July 3, 1931.
[73] "Annual Report of the Office of Motion Pictures," August 15, 1932.
[74] Ibid., 2.
[75] "Annual Report of the Office of Motion Pictures," August 15, 1932.
[76] Ibid., 2.
[77] Ibid., 3.
[78] Ibid., 5.
[79] Tugwell to Bradley, December 28,1934.
[80] Evans, "U.S.D.A. Motion Picture Service, 1908–1943," 7.
[81] "Annual Report, Motion Picture Service, Calendar Year 1938," n.d.
[82] Ibid.
[83] Ibid.
[84] "Supplementary Statement concerning United States Department of Agriculture Motion Picture Work," May 29,1939.
[85] Ibid.
[86] "Statement of the History and Work of the Motion Picture Section, U.S.D.A.," 4.
[87] "Report of Motion Picture Activities during the Month of June, 1942," July 14, 1942.
[88] Chester A. Lindstrom, "Agricultural Motion Pictures and the War," September 29, 1943, 9–10.
[89] Ibid.
[90] "Condensed Annual Report, Motion Picture Service, Fiscal Year 1946–1947," November 18, 1947.
[91] Ibid.

Chapter 2: *Helping Negroes to Become Better Farmers and Homemakers*

[1] Neyland, *Historically Black Land-Grant Institutions.*
[2] "Report to the Secretary of Agriculture on the Work of the Committee on Motion Picture Activities," November 30, 1914, 8.
[3] Kirby, "Black and White," 180.
[4] Conrad, *Forgotten Farmers,* 7.
[5] Ibid., 8.
[6] "Extension Work among Negroes, 1920," USDA Circular 190, 1921.
[7] Ibid.
[8] Ibid., 18–19.
[9] Ibid., 2.
[10] "Extension Work among Negroes, 1920."
[11] Kirby, "Black and White,"181.
[12] "Report to the Secretary of Agriculture on the Work of the Committee on Motion Picture Activities," November 30, 1914, 8.
[13] Conrad, *Forgotten Farmers,* 15.
[14] Tannous, "Extension Work," 94.
[15] Crosby, "Building the Country Home"; Neyland, *Historically Black Land-Grant Institutions.*
[16] Campbell, *Movable School.*
[17] A. Jones, "Role of Tuskegee Institute."
[18] Ibid.
[19] Crosby, "Building the Country Home," 28.
[20] Ibid., 33.
[21] Ibid., 34.
[22] Ibid., 36.
[23] Campbell, *Movable School.*
[24] Baker, *County Agent,* 202.
[25] See Bogle, *Toms, Coons, Mulattos*; Hall, "Whites of Their Eyes."
[26] Noble, *Negro in Films,* 11–12.
[27] Bogle, *Toms, Coons, Mulattos,* 7.
[28] Ibid., 8.
[29] Bogle, *Toms, Coons, Mulattos*; Hall, "Whites of Their Eyes"; Noble, *Negro in Films.*
[30] Rhodes, "Visibility of Race."
[31] Bogle, *Toms, Coons, Mulattos,* 8.
[32] "Annual Report of the Office of Motion Pictures," August 15, 1932, 3.
[33] "From the Report of the Editor," 4.
[34] *Birmingham Age Herald,* April 17, 1922, "Auburn's Extension Work among Negroes Shown in Film Tomorrow Night," 1.
[35] Ibid.
[36] "Excerpts from Letters in Regard to Motion Picture Films of the Department," November 24, 1922, 5.

[37] Crosby, "Building the Country Home," 51.
[38] Baker, *County Agent,* 203.
[39] Cox, "Alabama Farm Agents," 289.
[40] "For the Secretary," memorandum, April 25, 1919.
[41] See Packard, *American Nightmare,* 87.
[42] Conrad, *Forgotten Farmers,* 4.
[43] Simpson, "*Copycat,* Serial Murder," 148.
[44] Conrad, *Forgotten Farmers,* 8.
[45] Crosby, "Building the Country Home," 196.
[46] "Excerpts from Letters in Regard to Motion Picture Films of the Department," November 24, 1922, 5.
[47] Ibid., 6.
[48] Warburton to Secretary of State, October 23, 1929.
[49] "Annual Report of the Office of Motion Pictures," June 30, 1927.
[50] Warburton to Secretary of State, March 7, 1931.
[51] "Motion Pictures of the United States Department of Agriculture, 1934," catalog, 1934, 34
[52] Smith to Mr. Secretary, December 28, 1921.
[53] Martin to Council for National Defense, April 23, 1917.
[54] Virginia Agricultural and Industrial Institute to President of the United States, April 23, 1917.
[55] Jackson to Pearson, May 10, 1917 (included with p. 3 of the letter).
[56] Pearson to Jackson, May 16, 1917.
[57] Ousley to Campbell, November 1, 1918.
[58] "Extension Work among Negroes, 1920," USDA Circular 190, 1921, 5–6.
[59] Ibid.
[60] Myers, "Black Threat," 375.
[61] "Extension Work among Negroes, 1920," USDA Circular 190, 1921, 6.
[62] Meredith to Wasson, May 24, 1920.
[63] "Congressional Investigation of Lynching," NAACP press release, July 10, 1919.
[64] "Four Open Letters from the University Commission on Race Questions to the College Men of the South," August 31, 1917.
[65] James E. Gregg, "Lynching a National Menace, The White South's Protest against Lynching," January 5, 1916.
[66] Walton to Knapp, December 31, 1918, 2.
[67] "For the Secretary," memorandum, April 25, 1919.
[68] Blackshear to colleague, March 23, 1919.
[69] "Extension Work among Negroes, 1920," 5.

Chapter 3: *The Negro Farmer*

[1] "Annual Report of the Office of Motion Pictures," July 3, 1931.
[2] Harrison, *Black Exodus.*
[3] Packard, *American Nightmare,* 108.
[4] Grossman, "Black Labor," 57.

[5] Kirby, "Black and White."
[6] Packard, *American Nightmare*, 110.
[7] For example, see Wallace to Remmel, October 8, 1923.
[8] "Excerpts from Letters in Regard to Motion Picture Films of the Department," November 24, 1922, 5.
[9] Ibid., 6.
[10] "Annual Report of the Office of Motion Pictures," June 30, 1929, 2.
[11] "Annual Report of the Office of Motion Pictures," July 3, 1931.
[12] "Annual Report of the Office of Motion Pictures," August 15, 1932, 5.
[13] "Annual Report of the Office of Motion Pictures," June 30, 1927.
[14] See Chapter 2 for more on this topic.
[15] For an explanation and history of the 4-H clubs, see Reck, *4-H Story*. The USDA sponsored 4-H clubs as part of its extension work with the idea that teaching young people about agricultural advancements would encourage them to adopt them as adult farmers. The clubs also functioned as a way to encourage young people to be interested in agricultural pursuits.
[16] Washington, "Case of the Negro," 582.
[17] See Cohen, "Great Migration," and McMillen, "Migration and Black Protest."
[18] Grossman, "Black Labor," 55.
[19] For more, see Bailey, *Seaman A. Knapp*.
[20] See Chapter 2 for more on the difference between extension agents and home demonstration agents; briefly, extension agents were male while home demonstration agents were female, and their work was separated accordingly.
[21] Conrad, *Forgotten Farmers*, 15.
[22] Ibid., 8.
[23] "Informal Report on Motion Picture Work, June 1 to 30, 1918," July 1, 1918.
[24] Conrad, *Forgotten Farmers*, 7.
[25] The USDA's 4-H program is the focus of a later section of this chapter.
[26] Conrad, *Forgotten Farmers*, 8.
[27] Ibid., 16.
[28] Ibid.
[29] Ibid.,15.
[30] Washington, "Case of the Negro."
[31] Ibid., 577.
[32] Conrad, *Forgotten Farmers*, 4.
[33] See Bogle, *Toms, Coons, Mulattos,* for an explanation of these racist stereotypes.
[34] Hall, "Whites of Their Eyes," 36; emphasis in original.
[35] Conrad, *Forgotten Farmers*, 4–7.
[36] Hall to Moton, January 10, 1922.
[37] Grossman, "Black Labor," 55.
[38] See Chapter 2 for a full explanation.
[39] Higginbotham, *Shades of Freedom*.
[40] Washington, "Case of the Negro," 582.
[41] Jackson to Pearson, May 10, 1917, 3.
[42] Pearson to Jackson, May 16, 1917.
[43] "Extension Work among Negroes, 1920," USDA Circular 190, 1921, 5–6.

[44] Harris to Wallace, July 20, 1922.

[45] Remmel to Wallace, October 3, 1923.

[46] Wallace to Remmel, October 8, 1923. Note, in a subsequent letter, Remmel rescinds his request based on the objections of "a fraction of the colored people" in his state. Remmel to Wallace, October 11, 1923, Box 1, Folder 12: 1923, Records of the Dept. of Agriculture General Correspondence Relating to Negroes 1909–1955, RG 16, NARA.

[47] Jump to Wallace, October 11, 1923.

[48] "Memorandum for the Secretary," January 31, 1924, Box 2, Folder 1: 1924, Records of the Dept. of Agriculture General Correspondence Relating to Negroes 1909–1955, NARA.

[49] Conrad, *Forgotten Farmers*, 7.

[50] Packard, *American Nightmare*, 62.

[51] Conrad, *Forgotten Farmers*, 5.

[52] Grossman, "Black Labor."

Chapter 4: *Three Counties against Syphilis*

[1] "Motion Pictures of the United States Department of Agriculture, 1951," catalog, 1951, 44.

[2] J. Jones, *Bad Blood*, 43–4.

[3] Ibid., Foreword, xii.

[4] Ibid., 1.

[5] F. Gray, *Tuskegee Syphilis Study*, 14.

[6] Parran, "Shadow on the Land," 61.

[7] It should be noted that during the title sequence, one reference is made to the "traditional Negro music" being performed by the Hampton Institute Choir and Glee Club. This is the first indication that the film will address black culture.

[8] Savitt, *Race and Medicine*, 62.

[9] J. Jones, *Bad Blood*, 24, 43.

[10] Parran, "Shadow on the Land," 61.

[11] J. Jones, *Bad Blood*, 7.

[12] Brandt, "Racism and Research," 19.

[13] Rosenkrantz, "Non-Random Events," 242.

[14] The MPS had been shooting microscopic scenes since 1918. "Informal Report of Motion Picture Work, December 1 to 31, 1918," January 17, 1919.

[15] Additionally, the USDA filmmakers had a professional interest in developing different cinematic techniques for the presentation of scientific topics. The USDA archives reveal that since the earliest years, the USDA Motion Picture Service employees sought to develop their filmmaking along "experimental lines" to "make contributions to motion picture art." "To the Secretary of Agriculture on the Work of the Committee on Motion Picture Activities," memorandum, November 30, 1914, 5. The presentation of microscopic material was one area in which the USDA motion picture division was especially successful.

[16] Rosenkrantz, "Non-Random Events," 242.
[17] J. Jones, *Bad Blood*, 51.
[18] Ibid., 23.
[19] Lincoln and Mamiya, *Black Church*, 1990.
[20] Ibid., 17.
[21] Parran, "Shadow on the Land,"60.
[22] Ibid., 61.
[23] Ibid.
[24] Ibid., 60.
[25] J. Jones, *Bad Blood*, 73.
[26] Ibid.
[27] Ibid., 74.
[28] Lincoln and Mamiya, *Black Church*, 17.
[29] Hall, "Whites of Their Eyes," 28.
[30] Ibid., 42.
[31] Bogle, *Toms, Coons, Mulattos*, 4.
[32] Ibid.
[33] Hall, "Whites of Their Eyes," 33.
[34] J. Jones, *Bad Blood*, 16–20.
[35] Ibid., 17.
[36] Brandt, "Racism and Research," 16.
[37] Ibid., 17.
[38] J. Jones, *Bad Blood*, 18.
[39] Ibid., 24.
[40] Hall, "Whites of Their Eyes," 28.
[41] Ibid., 35.
[42] Ibid., 36, 37.
[43] Ibid., 37–8.
[44] F. Gray, *Tuskegee Syphilis Study*.
[45] J. Jones, *Bad Blood*, 162.
[46] "Motion Pictures of the United States Department of Agriculture, 1951," catalog, 1951, 44.
[47] Brandt, "Racism and Research," 15.

Chapter 5: *Henry Browne, Farmer*

[1] The film opens with a title explaining that it is presented by the U.S. Department of Agriculture. The opening titles also state that the motion picture is directed by Roger Barlow and that the music was composed by Gene Ferrell.
[2] For an in-depth treatment of Jim Crow laws, see Packard, *American Nightmare*.
[3] Canada Lee is credited in the opening titles as the narrator. For more on Canada Lee, see Smith, *Becoming Something*.
[4] Conkin, *Revolution Down on the Farm*, 62.
[5] Conrad, *Forgotten Farmers*.
[6] Hurt, *American Agriculture*, 288.

[7] Perkins, *Crisis in Agriculture*, 7.
[8] Ibid., 194.
[9] Conrad, *Forgotten Farmers*, 205.
[10] Auerbach, "Southern Tenant Farmers," 56–7.
[11] One white soldier is seen briefly at the beginning of the Tuskegee Airmen sequence.
[12] The Browne farmhouse is a dramatic break from the rundown shacks presented as African American farmhouses in *The Negro Farmer*; the Browne home appears to be soundly built and well kept.
[13] Noble, *Negro in Films.*
[14] Francis, *Tuskegee Airmen.*
[15] The 99th Squadron became the first all-black squadron to enter combat in World War II; it arrived in North Africa on May 31, 1943, a year after the film was completed.
[16] Cripps, *Slow Fade to Black*, 379.
[17] Noble, *Negro in Films*, 106.
[18] Cripps, *Slow Fade to Black*, 379.
[19] Cripps and Culbert, "Negro Soldier," 112.
[20] "To Chiefs of Bureaus and Independent Offices," memorandum, March 24, 1914.
[21] Ibid.
[22] Noble, *Negro in Films*, 106.
[23] Washington, "The Case of the Negro."
[24] Ibid., 582.
[25] Ibid., 578.
[26] As I have pointed out earlier in this book, the doctrine of separate but equal was a lie, as African Americans were institutionally separated from whites but were never provided equal treatment. See Packard, *American Nightmare*, 87.
[27] Ibid., 108.
[28] "Agricultural Motion Pictures and the War," 6.
[29] Hamilton to Goss, March 11, 1943.
[30] "Field Reports on the Use of Department of Agriculture Motion Pictures," 2.
[31] "Field Reports on the Use of Department of Agriculture Motion Pictures," 9.
[32] White to Wickard, November3, 1942.
[33] Cripps and Culbert, "Negro Soldier," 112.

Chapter 6: USDA Motion Picture Representations of African Americans

[1] Packard, *American Nightmare*, 62.
[2] Ibid., 77.
[3] Rhodes, "Visibility of Race," 187.
[4] Packard, *American Nightmare*, 87.
[5] Bogle, *Toms, Coons, Mulattos*; Rhodes, "Visibility of Race"; Hall, "Whites of Their Eyes."

[6] Bogle, *Toms, Coons, Mulattos,* 8.
[7] Washington, "Case of the Negro," 577.
[8] Myers, "Black Threat," 375.
[9] Conrad, *Forgotten Farmers,* 10.
[10] Ibid., 4.
[11] Myers, "Black Threat," 375.
[12] Washington, "Case of the Negro," 582; quoted in *The Negro Farmer,* 1938, USDA motion picture.
[13] Jump to Wallace, October 11, 1923.
[14] "Memorandum for the Secretary," January 31, 1924, Box 2, Folder 1: 1924, Records of the Dept. of Agriculture General Correspondence Relating to Negroes 1909–1955, NARA.
[15] Wilkerson, *Warmth of Other Suns,* 9–10.
[16] Ibid., 10.
[17] See J. Jones, *Bad Blood.*
[18] White to Wickard, November 3, 1942.
[19] Myers, "Black Threat," 373–4.
[20] Hall, "Whites of Their Eyes," 33.
[21] Bogle, *Toms, Coons, Mulattos,* 4.
[22] Rhodes, "Visibility of Race," 186.
[23] Higginbotham, *Shades of Freedom,* 9.
[24] Packard, *American Nightmare,* 86.
[25] Myers, "Black Threat," 375.
[26] Higginbotham, *Shades of Freedom,* 120.
[27] Rhodes, "Visibility of Race," 185.
[28] Hall, "Whites of Their Eyes," 46.
[29] Myers, "Black Threat," 375.
[30] Higginbotham, *Shades of Freedom,* 3–4.
[31] Ibid., 14.
[32] Ibid., 41. Higginbotham explains, "To be white meant that one did not have any black blood."
[33] Packard, American Nightmare, 62.
[34] Ibid., 73.
[35] See Hall, "Whites of Their Eyes."
[36] Ibid., 36.
[37] Higginbotham, *Shades of Freedom.*
[38] Kellner, "Cultural Studies," 15.
[39] Kellner and Share, "Toward Critical Media Literacy," 372.
[40] Ibid., 370.
[41] Hall, "Whites of Their Eyes," 37.
[42] Kellner, "Multiple Literacies," 200.
[43] Ibid., 201.

Bibliography

Aden, R. C. "Back to the Garden: Therapeutic Place Metaphor in *Field of Dreams.*" *Southern Journal of Communication* 59 (1994): 307–17.

"Auburn's Extension Work among Negroes Shown in Film Tomorrow Night," *Birmingham Age Herald,* April 17, 1922.

Auerbach, Jerold S. "Southern Tenant Farmers: Socialist Critics of the New Deal." In Dubofsky and Burwood, eds., *Agriculture during the Great Depression,* 55–70.

Bailey, Joseph C. *Seaman A. Knapp, School Master of American Agriculture.* New York: Columbia University Press, 1945.

Baker, Gladys L. *The County Agent.* Chicago: University of Chicago Press, 1939.

Barnouw, Erik. *Documentary: A History of the Non-Fiction Film.* Rev. ed. Oxford: Oxford University Press, 1983.

Benson, Thomas W. "Looking for the Public in the Popular: The Hollywood Blacklist and the Rhetoric of Collective Memory." In Blakesley, ed., *Terministic Screen,* 129–45.

Blakesley, David, ed. *The Terministic Screen: Rhetorical Perspectives on Film.* Carbondale: Southern Illinois University Press, 2003.

Bobo, J. *Black Women as Cultural Readers.* New York: Columbia University Press, 1995.

Bogle, Donald. *Toms, Coons, Mulattos, Mammies, and Bucks: An Interpretive History of Blacks in American Films.* 3rd edn. New York: Continuum, 1996.

Brandt, Allan M. "Racism and Research: The Case of the Tuskegee Syphilis Experiment." In Reverby, *Tuskegee's Truths,* 15–33.

Bridges, George, and Rosalind Brunt, eds. *Silver Linings: Some Strategies for the Eighties.* London: Lawrence and Wishart, 1981.

Brinson, Susan L. "The Myth of White Superiority in *Mississippi Burning.*" *Southern Communication Journal* 60 (1995): 211–21.

Brummett, Barry. "Electric Literature as Equipment for Living: Haunted House Films." *Critical Studies in Mass Communication* 2 (1985): 247–61.

Campbell, Thomas M. *The Movable School Goes to the Negro Farmer.* New York: Arno Press, 1969.

Carey, James. *Communication as Culture.* Boston: Unwin Hyman, 1989.

Cohen, William. "The Great Migration as a Lever for Social Change." In Harrison, *Black Exodus,* 72–82.

Conkin, Paul K. *A Revolution Down on the Farm: The Transformation of American Agriculture since 1929.* Lexington: University Press of Kentucky, 2008.

Conrad, David Eugene. *The Forgotten Farmers: The Story of Sharecroppers in the New Deal.* Urbana: University of Illinois Press, 1965.

Cox, Dewayne. "Alabama Farm Agents, 1914–1922." *Alabama Review* 47 (1994): 285–304.

Cripps, Thomas. *Slow Fade to Black: The Negro in American Film, 1900–1942.* New York: Oxford University Press, 1977.

Cripps, Thomas, and David Culbert. "*The Negro Soldier* (1944): Film Propaganda in Black and White." In Rollins, *Hollywood as Historian,* 109–33.

Crosby, Earl William. "Building the Country Home: The Black County Agent System, 1906–1940." PhD dissertation, Miami University, 1977.

Diawara, M. (1988). "Black Spectatorship: Problems of Identification and Resistance." *Screen* 29 (1988): 66–76.

Dubofsky, Melvin, and Stephen Burwood, eds. *Agriculture during the Great Depression.* New York: Garland, 1990.

Ellis, Don Carlos, and Laura Thornborough. *Motion Pictures in Education: A Practical Handbook for Users of Visual Aids.* New York: Cromwell, 1923.

Ellis, Jack C., and Betsy A. McLane. *A New History of Documentary Film.* New York: Continuum, 2009.

Francis, Charles E., *The Tuskegee Airmen: The Men Who Changed a Nation.* Boston: Braden, 1993.

Frentz, T. S. and T. B. Farrell. "Conversion of America's Consciousness: The Rhetoric of *The Exorcist.*" *Quarterly Journal of Speech* 61 (1975): 40–7.

Frentz, T. S., and M. E. Hale. "Inferential Model of Criticism of *The Empire Strikes Back.*" *Quarterly Journal of Speech* 69 (1983): 278–89.

Frentz, T. S. and J. Rushing. "The Rhetoric of *Rocky: Part Two.*" *Western Journal of Speech Communication* 42 (1978): 231–40.

Gray, Fred D. *The Tuskegee Syphilis Study: The Real Story and Beyond.* Montgomery, AL: Black Belt Press, 1998.

Gray, Herman. "Television, Black Americans, and the American Dream." *Critical Studies in Mass Communication* 6 (1989): 376–86.

Grossman, James R. "Black Labor Is the Best Labor: Southern White Reactions to the Great Migration." In Harrison, *Black Exodus,* 51–71.

Hall, Stuart. "Culture, the Media, and the 'Ideological Effect.'" In *Mass Communication and Society,* edited by J. Curran, M. Gurevitch, and J. Wollacott, 315–48. Beverly Hills, CA: Sage, 1979.

—. "The Rediscovery of Ideology: Return of the Repressed in Media Studies." In *Culture, Society, and the Media,* edited by M. Gurevitch, T. Bennett, J. Curran, and J. Woollocott, 56–91. London: Methuen, 1982.

—. "The Whites of Their Eyes: Racist Ideologies and the Media." In Bridges and Brunt, *Silver Linings,* 28–52.

Harrison, Alferdteen, ed. *Black Exodus: The Great Migration from the American South.* Jackson: University Press of Mississippi, 1991.

Higginbotham, A. Leon, Jr. *Shades of Freedom: Racial Politics and Presumptions of the American Legal Process.* New York: Oxford University Press, 1996.

Hurt, R. Douglas. *American Agriculture: A Brief History*. Ames: Iowa State University Press, 1994.

Jameson, Fredric. *The Political Unconscious: Narrative as a Socially Symbolic Act.* Ithaca, NY: Cornell University Press, 1981.

Jones, Allen W. "The Role of Tuskegee Institute in the Education of Black Farmers." *Journal of Negro History* 60, no. 2 (1975): 252–67.

Jones, James H. *Bad Blood: The Tuskegee Syphilis Experiment.* New York: Free Press, 1993.

—. Foreword to *Tuskegee's Truths.* In Reverby, xii.

Kellner, Douglas. "Cultural Studies, Multiculturalism and Media Culture." In *Gender, Race and Class in Media: A Text-Reader,* edited by Gail Dines and J. M. Humez, 5–17. Thousand Oaks, CA: Sage, 1995.

—. "Multiple Literacies and Critical Pedagogies: New Paradigms." In *Revolutionary Pedagogies: Cultural Politics, Instituting Education, and the Discourse of Theory,* edited by Peter Pericles Trifonas, 196–221. New York: RoutledgeFalmer, 2000.

Kellner, Douglas, and Jeff Share. "Toward Critical Media Literacy: Core Concepts, Debates, Organizations, and Policy." *Discourse: Studies in the Cultural Politics of Education* 26, no. 3 (2005): 369–86.

Kirby, Jack Temple. "Black and White in the Rural South, 1915–1954." In Dubofsky and Burwood, *Agriculture during the Great Depression,* 179–90.

Lincoln, C. Eric, and Lawrence H. Mamiya. *The Black Church in the African American Experience.* Durham, NC: Duke University Press, 1990.

MacCann, Richard Dyer. *The People's Films: A Political History of U.S. Government Motion Pictures.* New York: Hastings House, 1973.

McMillen, Neil. "The Migration and Black Protest in Jim Crow Mississippi." In Harrison, *Black Exodus,* 83–99.

Myers, Martha A. "Black Threat and Incarceration in Postbellum Georgia." *Social Forces* 69(2) (1990): 373–93.

Neyland, Leedell W. *Historically Black Land-Grant Institutions and the Development of Agriculture and Home Economics, 1890–1990.* Tallahassee: Florida A&M University Foundation, 1990.

Nichols, Bill. *Representing Reality: Issues and Concepts in Documentary.* Bloomington: Indiana University Press, 1991.

Noble, Peter. *The Negro in Films.* London: Skelton Robinson, 1948.

Packard, Jerrold M. *American Nightmare: The History of Jim Crow.* New York: St. Martin's Press, 2002.

Parran, Thomas. "Shadow on the Land: Syphilis, the White Man's Burden." In Reverby, *Tuskegee's Truths,* 59–69.

Payne, D. "Characterology, Media, and Rhetoric." In *Media, Consciousness, and Culture,* edited by Bruce E. Gronbeck, Thomas J. Farrell, and Paul A. Soukup. Thousand Oaks, CA: Sage, 1991: 223–36.

—. "Political Vertigo in *Dead Poets Society.*" *Southern Communication Journal* 58 (1992): 13–21.

—. "*The Wizard of Oz:* Therapeutic Rhetoric in a Contemporary Ritual." *Quarterly Journal of Speech* 75 (1989): 25–39.

Perkins, Van L. *Crisis in Agriculture: The Agricultural Adjustment Administration and the New Deal, 1933.* Berkeley: University of California Press, 1969.

Rasmussen, K., and S. D. Downey. "Dialectical Disorientation in *Agnes of God*." *Western Journal of Speech Communication* 53 (1989): 66–84.

—. "Dialectical Disorientation in Vietnam War Films: Subversion of the Mythology of War." *Quarterly Journal of Speech* 77 (1991): 176–95.

Reck, Franklin M. *The 4-H Story: A History of 4-H Club Work*. Ames: Iowa State College Press, 1951.

Reverby, Susan M., ed. *Tuskegee's Truths: Rethinking the Tuskegee Syphilis Study*. Chapel Hill: University of North Carolina Press, 2000.

Rhodes, Jane. "The Visibility of Race and Media History." *Critical Studies in Mass Communication* 10 (1993): 184–90.

Rollins, Peter C., ed. *Hollywood as Historian: American Film in a Cultural Context*. Rev. ed. Lexington: University Press of Kentucky, 1983.

Rosenkrantz, Barbara. "Non-Random Events." In Reverby, *Tuskegee's Truths*, 236–47.

Ross, Steven. J. "Struggles for the Screen: Workers, Radicals, and the Political Uses of Silent Film." *American Historical Review* 96 (1991): 333–67.

Rushing, J. H. "Evolution of 'the New Frontier' in *Alien* and *Aliens:* Patriarchal Co-optation of the Feminine Archetype." *Quarterly Journal of Speech* 75 (1989): 1–24.

—. "Frontierism and the Materialization of the Psyche: The Rhetoric of *Innerspace*." *Southern Communication Journal* 56 (1991): 243–56.

—. "Mythic Evolution of 'the New Frontier' in Mass Mediated Rhetoric." *Critical Studies in Mass Communication* 3 (1986): 265–96.

—. "The Rhetoric of the American Western Myth." *Communication Monographs* 50 (1983): 14–32.

Rushing, J. H., and T. S. Frentz. "*The Deer Hunter*: Rhetoric of the Warrior." *Quarterly Journal of Speech* 66 (1980): 392–406.

—. "The Frankenstein Myth in Contemporary Cinema." *Critical Studies in Mass Communication* 6 (1989): 61–80.

—. "The Rhetoric of *Rocky*: A Social Value Model of Criticism." *Western Journal of Speech Communication* 41 (1978): 63–72.

Savitt, Todd L. *Race and Medicine in Nineteenth- and Early Twentieth-Century America*. Kent, OH: Kent State University Press, 2007.

Simpson, P. L. "*Copycat,* Serial Murder, and the (De)terministic Screen Narrative." In Blakesley, *Terministic Screen,*146–62.

Smith, Mona Z. *Becoming Something: The Story of Canada Lee*. New York: Faber and Faber, 2004.

Solomon, M., and W. J. McMullen. "*Places in the Heart:* The Rhetorical Force of an Open Text." *Western Journal of Speech Communication* 55 (1991): 339–53.

Tannous, A. I. "Extension Work among the Arab Fellahin." In *Farmers of the World: The Development of Agricultural Extension,* edited by E. Brunner, I. T. Sanders, and D. Ensminger, 78–100. New York: Columbia University Press, 1945.

Washington, Booker T. "The Case of the Negro." *Atlantic Monthly,* 84 (1899): 577–87.

Wilkerson, Isabel. *The Warmth of Other Suns: The Epic Story of America's Great Migration*. New York: Random House, 2010.

Winn, J. Emmett. *The American Dream and Contemporary Hollywood Cinema.* New York: Continuum, 2007.
—. "Documenting Racism in an Agricultural Extension Film." *Film and History* 38, no. 1 (2008): 33–43.

Archival Sources

Unless otherwise specified, all archival sources listed are from Record Group 16 of the National Archives and Records Administration, College Park, Maryland. Abbreviations used:

RDA—Records of the Dept. of Agriculture General Correspondence Relating to Negroes, 1909–1955.
HMPS—Records Relating to the History of the Motion Picture Service, 1908–1958.
ROSA—Records of the Office of the Secretary of Agriculture, General Correspondence of the Office of the Secretary, 1906–1970.

"Annual Report, Motion Picture Service, Calendar Year 1938," n.d., Box 1, Folder: Annual Reports 1937–1938. HMPS.
"Annual Report of the Office of Motion Pictures, Extension Service, U.S. Department of Agriculture, Fiscal Year Ending June 30, 1919," June 30, 1919, Box 1, Folder: Annual. HMPS.
"Annual Report of the Office of Motion Pictures, Extension Service, U.S. Department of Agriculture, Fiscal Year Ending June 30, 1923," June 30, 1923, Box 1, Folder: Annual. HMPS.
"Annual Report of the Office of Motion Pictures, Extension Service, U.S. Department of Agriculture, Fiscal Year Ending June 30, 1927," June 30, 1927, Box 1, Folder: Annual. HMPS.
"Annual Report of the Office of Motion Pictures, Extension Service, U.S. Department of Agriculture, Fiscal Year Ending June 30, 1928," June 30, 1928, Box 1, Folder: Annual. HMPS.
"Annual Report of the Office of Motion Pictures, Extension Service, U.S. Department of Agriculture, Fiscal Year Ending June 30, 1929," June 30, 1929, 2, Box 1, Folder: Annual. HMPS.
"Annual Report of the Office of Motion Pictures, Extension Service, U.S. Department of Agriculture, Fiscal Year Ending June 30, 1931," July 3, 1931, Box 1, Folder: Annual. HMPS.
"Annual Report of the Office of Motion Pictures, Extension Service, U.S. Department of Agriculture, Fiscal Year Ending June 30, 1932," August 15, 1932, Box 1, Folder: USDA MPS Facilities. HMPS.
Ball, Eustace Hale, to David Franklin Huston, April 2, 1913, Box 56, Folder: Moving Pictures 1912–1913. ROSA.
Blackshear, F. L., to colleague, March 23, 1919, Box 1, Folder 8: 1919. RDA
"Condensed Annual Report, Motion Picture Service, Fiscal Year 1946–1947," November 18, 1947, Box 1, Folder: Reports 7 Annual. HMPS.
"Congressional Investigation of Lynching," NAACP press release, July 10, 1919, Box 1, Folder 8: 1919. RDA.

Evans, Raymond. "The U.S.D.A. Motion Picture Service, 1908–1943," n.d. (probably 1943), Box 5, Folder: Educational Screen, Nov. 1937. HMPS.
"Excerpts from Letters in Regard to Motion Picture Films of the Department," November 24, 1922, Box 915, Folder: Moving Pictures. ROSA.
"Extension Work among Negroes, 1920," USDA Circular 190, 1921, Box 1, Folder 10: 1921. RDA.
"Farm Films Seen around the World," newspaper clipping from LaGrange, Georgia, March 8, 1929, Box 1437, Folder: Motion Pictures Jan.–June. ROSA.
"Field Reports on the Use of Department of Agriculture Motion Pictures," n.d. (probably 1943), Box 4, Folder: Field Reports 1943. HMPS.
"For Dr. Galloway," memorandum, January 6, 1914, Box 151, 1914. ROSA.
"For Dr. Galloway," memorandum, February 16, 1914, Box 151, Folder: 1914. ROSA.
"For Dr. Warburton, Director of Extension Work," memorandum, April 29, 1930, Box 2, Folder: Articles for Extension Review, 3-22-30 to 5-10-30. HMPS.
"For Heads of Bureaus, Offices, and Independent Divisions," memorandum, May 24, 1913, Box 56, Folder: Moving Pictures 1912–1913. ROSA
"For Mr. G. W. Wharton," memorandum, February 4, 1914, Box 151, Folder: 1914. ROSA.
"For the Secretary," memorandum, July 7, 1917, Box 415, Folder: Moving Pictures 1917. ROSA.
"For the Secretary," memorandum, April 25, 1919, Box 1, Folder 8: 1919. RDA.
"Four Open Letters from the University Commission on Race Questions to the College Men of the South," August 31, 1917, Box 1, Folder 8: 1919. RDA.
"From the Report of the Editor, Department of Agriculture, 1913–1916," n.d., Box 2, Folder: Motion Picture Service. HMPS.
Gregg, James E. "Lynching a National Menace: The White South's Protest against Lynching," January 5, 1916. Pamphlet published by Hampton Institute, Box 1, Folder 8: 1919. RDA.
Hall, Charles E., to Robert R. Moton, January 10, 1922, Box 1, Folder 11: 1922. RDA.
Hamilton, Carl, to A. S. Goss, March 11, 1943, Box 855, Folder: Illustrations 2–3 Motion Pictures, Lantern Slides, Film Strips. ROSA.
Harris, G. Silas, to Henry C. Wallace, July 10, 1922, Box 1, Folder 11: 1922. ROSA.
"Informal Report on Motion Picture Work, March 1 to 15," March 8, 1918, Box 578, Folder: Moving Pictures 1918. ROSA.
"Informal Report on Motion Picture Work, April 1 to 15, 1918,", April 15, 1918, Box 578, Folder: Moving Pictures 1918. ROSA.
"Informal Report on Motion Picture Work, June 1 to 30, 1918," July 1, 1918, Box 578, Folder: Moving Pictures 1918. ROSA.
"Informal Report on Motion Picture Work, December 1 to 31, 1918," January 17, 1919, Box 683, Folder: Moving Pictures 1919. ROSA.
"Informal Report of Motion Picture Work, April and May, 1919," n.d., Box 683, Folder: Moving Pictures 1919. ROSA.
Jackson, Giles B., to R. A. Pearson, May 10, 1917, 3, Box 1, Folder 6: 1917. RDA.
Jump, W. A., to Henry C. Wallace, October 11, 1923, Box 1, Folder 12: 1923. RDA.

Lindstrom, Chester A. "Agricultural Motion Pictures and the War," September 29, 1943, Box 3, Folder: Paper prepared by Chester A. Lindstrom; delivered at the 54th semiannual Engineering Conference. HMPS.

—. "Films Working for Agriculture," April 18, 1945, 2, Box 3, Folder: Films Working for Agriculture. HMPS.

Martin, Thomas S., to Council for National Defense, April 23, 1917, Box 1, Folder 6: 1917. RDA.

Meredith, E., to R. C. Wasson, May 24, 1920, Box 1, Folder 9: 1920, 170-10-25. RDA.

"Motion Pictures of the United States Department of Agriculture, 1934," catalog, 1934, Box 5, Folder: Motion Picture Catalogs, HMPS.

"Motion Pictures of the United States Department of Agriculture, 1951," catalog, 1951, Box 5, Folder: Motion Picture Catalogs. HMPS.

"Motion Picture Policies of the U.S. Department of Agriculture," January 31, 1923, Box 998, Folder: Moving Pictures Jan. 1–July 31. ROSA.

"No. 55, Regarding Motion Picture Activities of the Department," memorandum, December 18, 1913, Box 299, Folder: Moving Pictures. ROSA.

"No. 85, The Purchase of Motion-Picture Apparatus," memorandum, April 20, 1914, Box 299, Folder: Moving Pictures. ROSA.

Ousley, Clarence, to William Bacon, July 25, 1917, Box 415, Folder: Moving Pictures 1917. ROSA.

Ousley, Clarence, to Thomas M. Campbell, November 1, 1918, Box 1, Folder 7: 1918. RDA.

Ousley, Clarence, to Carl Lemmle, December 5, 1917, Box 415, Folder: Moving Pictures 1917. ROSA.

Pearson, R. A., to Giles B. Jackson, May 16, 1917, Box 1, Folder 6: 1917. RDA.

Perkins, Fred W. "A Twelve-Year Trial of Educational Films," Reprinted from Transactions of the Society of Motion Picture Engineers, Number 26, 1926, 51, Box 2, Folder: History of Motion Pictures. HMPS.

Remmel, H. L., to Henry C. Wallace, October 3, 1923, Box 1, Folder 12: 1923. RDA.

—. October 11, 1923, Box 1, Folder 12: 1923. RDA.

"Report of Motion Picture Activities during the Month of June, 1942," July 14, 1942, Box 5, Folder: Report on Defense Activities Monthly. HMPS.

"Report to the Secretary of Agriculture on the Work of the Committee on Motion Picture Activities," November 30, 1914, Box 2, Folder: Report on Work of the Committee on Motion Picture Activities. HMPS.

Smith, Harlan, to Mr. Secretary, December 28, 1921, Box 915, Folder: Moving Pictures. ROSA.

"A Statement of the History and Work of the Motion Picture Section, U.S.D.A.," n.d. (probably 1941), Box 2, Folder: History of Department Activities as They Related to Defense. HMPS.

"Supplementary Statement concerning United States Department of Agriculture Motion Picture Work," May 29, 1939, Box 3048M, Folder: Motion Pictures. ROSA.

"To Chiefs of Bureaus and Independent Offices, Members of the Motion Picture Committee and Chiefs of Laboratories and Divisions Directly Affected," memorandum, March 24, 1914, Box 151, Folder: 1914. ROSA.

"To the Secretary of Agriculture on the Work of the Committee on Motion Picture Activities," memorandum, November 30, 1914, Box 151, Folder: 1914. ROSA.

Tugwell, R. G., to John G. Bradley, December 28, 1934, Box 2032, Folder: Motion Pictures. ROSA.

Untitled report, n.d., Box 2, Folder: History of Motion Pictures. HMPS.

U.S. Department of Agriculture. *Helping Negroes to Become Better Farmers and Homemakers.* 1921. Motion picture.

—. *Henry Browne, Farmer.* 1942. Motion picture.

—. *An International Study of American Roads.* Washington, DC: National Archives and Records Administration, 1932. Motion picture.

—. *The Negro Farmer.* 1938. Motion picture.

—. *Three Counties against Syphilis.* 1938. Motion picture.

"The Use of Educational Motion Pictures by the United States Department of Agriculture," September 1929, Box 2, Folder: Annual Reports of MP. HMPS.

"U.S. Takes Lead in Educational Film Production," *New York Herald Tribune,* Sunday, April 1, 1928, Box 1360, Folder: Motion Pictures Jan.–May. ROSA.

Virginia Agriculture and Industrial Institute to President of the United States, April 23, 1917, Box 1, Folder 6: 1917. RDA.

Wallace, Henry C.,to H. L. Remmel, October 8, 1923, Box 1, Folder 12: 1923. RDA.

Walton, T. O., to Bradford Knapp, December 31, 1918, Box 1, Folder 7: 1918. RDA.

Warburton, C. W., to Secretary of State, October 23, 1929, Box 1437, Folder Motion Pictures July-. ROSA.

Warburton, C. W., to Secretary of State, March 7, 1931, Box 1622, Folder: Motion Pictures (1 of 2). ROSA.

Warburton, C. W., to A. F. Woods, March 7, 1931, Box 1622, Folder: Motion Pictures (1 of 2). ROSA.

Warburton, C. W., "A World Extension Course Conducted through U.S. Films," February 17, 1930, Box 2, Folder: A World Extension Course. HMPS.

White, Walter, to Claude R. Wickard, November 3, 1942, Box 656, Folder: Illustrations 2 to 3 Motion Pictures. ROSA.

Index